Mathematics
Problem Solving
COACH

Strategies and Applications

G

Doreen S. Nation
Sheila J. Siderman

Mathematics Problem Solving Coach Level G: Strategies and Applications
67NA
ISBN# **1-58620-543-9**

EVP, Publisher: Steven Zweig
Managing Editor: Kelly Bellini
Creative Director: Spencer Brinker
Art Director: Farzana Razak
Authors: Siderman Nation Publishing Services, Inc.
　　　　Doreen S. Nation and Shiela J. Siderman
Contributing Editors: Constance Lehmann, Dena Lindsay
Interior Design: Rocio Paez
Cover Design: Farzana Razak
Production Manager: Michelle McGuinness
Cover Photo: Photodisc Green/Getty Images

Triumph Learning 333 East 38th Street, New York, NY 10016-2777
© 2004 Triumph Learning, LLC
A Haights Cross Communications company

All rights reserved. No part of this publication may be reproduced in whole or in part, stored in a retrieval system, or transmitted in any form or by any means, electronic, mechanical, photocopying, recording or otherwise, without written permission from the publisher.

Printed in the United States of America.

10 9 8 7 6 5 4 3 2

Mathematics Problem Solving Coach, Level G

Table of Contents

To the Teacher 5
To the Student 6
Problem-Solving Strategies 7

I. Strategies for Solving Problems

Lesson 1	Draw a Diagram 8
Lesson 2	Make a Model 10
Lesson 3	Make an Organized List 12
Lesson 4	Make a Table or Graph 14
Lesson 5	Look for a Pattern 16
Lesson 6	Predict and Test 18
Lesson 7	Use Logical Thinking 20
Lesson 8	Work Backward 22
Lesson 9	Solve a Simpler Problem 24
Lesson 10	Write a Formula or an Equation 26

II. Applying the Strategies

Solving Problems with Rational Numbers

Lesson 11	Operations with Whole Numbers, Fractions, and Decimals 28
Lesson 12	Operations with Integers 32
Lesson 13	Operations with Rational Numbers 36
Lesson 14	Number Theory 40
Lesson 15	Ratios, Rates, and Proportions 44
Lesson 16	Operations with Percents 48

NOTICE: Photocopying any part of this book is forbidden by law.

Solving Problems with Geometry and Measurement

Lesson 17	Metric and Customary Units of Measurement	52
Lesson 18	Lines and Angles	56
Lesson 19	Polygons and Circles	60
Lesson 20	Congruent and Similar Figures	64
Lesson 21	Rotations, Reflections, and Translations	68
Lesson 22	Solid Figures	72

Solving Problems with Algebra

Lesson 23	Compare and Order Rational and Irrational Numbers	76
Lesson 24	Expressions, One-Step, and Two-Step Equations	80
Lesson 25	Linear Equations	84

Solving Problems with Data and Probability

Lesson 26	Measures of Central Tendency	88
Lesson 27	Combinations and Permutations	92
Lesson 28	Experimental and Theoretical Probability	96

III. Practice Test

Answer Sheet	100
Test-Taking Tips	101
Practice Test	102

To the Teacher

Mathematics Problem Solving Coach has been created to help your seventh-grade students improve their performance on your state proficiency tests.

The first part of the book focuses on ten important problem-solving strategies:

Strategy 1:	Draw a Diagram
Strategy 2:	Make a Model
Strategy 3:	Make an Organized List
Strategy 4:	Make a Table or Graph
Strategy 5:	Look for a Pattern
Strategy 6:	Predict and Test
Strategy 7:	Use Logical Thinking
Strategy 8:	Work Backward
Strategy 9:	Solve a Simpler Problem
Strategy 10:	Write a Formula or an Equation

The second part of the book focuses on using these problem-solving strategies to solve problems like those your students will have to solve on the state test. These problems cover the range of seventh-grade math topics, including rational numbers, geometry, measurement, algebra, data, and probability.

The third part of the book provides test-taking tips, an eleven-page practice test, and an answer sheet. You may wish to have students use the answer sheet to practice how to record answers.

Most teachers use *Mathematics Problem Solving Coach* in one of two ways:

- Have students proceed straight through Lessons 1 to 28. Then give students the Practice Test.

- Begin by teaching the first ten lessons and then use the remaining lessons in conjunction with classroom study of that topic. Some students may need to use only those lessons in which they need further work. Then give students the Practice Test.

We hope and expect that by using this book, students will not only enhance their math competencies and do better on the state proficiency tests, but also that they will enjoy the problem-solving situations chosen.

NOTICE: Photocopying any part of this book is forbidden by law.

To the Student

This book, **Mathematics Problem Solving Coach**, will help you learn how to solve word problems. It will prepare you for your state math test.

This book contains these parts:

> **Part I   Strategies for Solving Problems**
>
> In this part of the book, you will learn 10 strategies that will help you solve problems. Each strategy shows you how to use the four-step problem-solving method:
> - Find Out
> - Plan
> - Solve
> - Look Back
>
> **Part II   Applying the Strategies**
>
> In this part of the book, you will practice using the 10 strategies as you solve problems. The problems include rational numbers, geometry, measurement, algebra, data, and probability.
>
> **Part III   Practice Test**
>
> In this part of the book, you will find tips for doing well on tests. You will take a test to practice your test-taking skills.

Mathematics Problem Solving Coach contains multiple-choice questions and open-ended questions. In the multiple-choice questions, there is only one correct answer. In the open-ended questions, you must write your own answer. For some questions, you also need to explain how you found your answer. For other questions, you will record your answer in a bubble grid.

Here is how you can use this book to prepare for your test:

1. In each lesson, read the Examples several times. This will help you see the steps needed to get the correct answer. If your class has another math book as well, you may want to find lessons in it that are like the Examples.

2. Try the sample test questions at the end of each lesson. If you cannot do some of them, study the Examples again.

3. As you work on Lessons 11–28, look back at Lessons 1–10 for problem-solving strategies you can use.

Good luck!

Ten Problem-Solving Strategies You Can Use To Be a Better Problem Solver

1 Draw a Diagram
You can use this strategy when you need to see the information given in a problem to be able to solve it.

2 Make a Model
You can use this strategy when you need to see the data. Then you can watch how the solution is found.

3 Make an Organized List
You can use this strategy to help you review the information in a problem. It will help you organize your thinking about it.

4 Make a Table or Graph
You can use this strategy to organize information given in a problem. Tables can help you see relationships and patterns in data.

5 Look for a Pattern
You can use this strategy to predict what comes next. If you can find the rule for a pattern, you can use it to solve the problem.

6 Predict and Test
You can use this strategy when it is difficult to work out the answer to a problem. You can make a guess and then test it. If your guess is not correct, use that guess to make a better guess.

7 Use Logical Thinking
You can use this strategy when you need to think about how the information you know fits together.

8 Work Backward
You can use this strategy when you know the total but need to find a missing part. Start at the end and work back to find the answer.

9 Solve a Simpler Problem
You can use this strategy when the numbers in a problem are very big. Change them to smaller numbers that are easier to work with. Decide how to solve the problem. Then solve the original problem the same way.

10 Write a Formula or an Equation
You can use this strategy when you need to find a missing amount to solve a problem.

Part I: Strategies for Solving Problems

1 Draw a Diagram

Sometimes organizing information in a diagram helps you solve a problem. You can use the strategy *draw a diagram*.

Example

The parks department is putting in two new public tennis courts. The courts will be side by side, with 18 feet between them. Each court is 36 feet wide by 78 feet long, with an extra 21 feet at each end and an extra 12 feet at each of the outer sides. How much fencing will the parks department need to enclose both courts?

Find Out

Think: What facts do you know?
- The courts will be side by side with 18 feet between them.
- Each court is 36 feet wide by 78 feet long.
- Each court has an extra 21 feet at each end and an extra 12 feet at each of the outer sides.

Think: What do you need to find out?
You want to find out how much fencing will be needed to enclose both courts.

Plan

Think: What strategy can you use?
You can draw a diagram of the tennis courts to find the perimeter of the courts.

Solve

Width = 12 + 36 + 18 + 36 + 12
= 114 feet

Length = 78 + 21 + 21 = 120 feet

Perimeter = 2 × (length + width)
= 2 × (120 + 114) = 2 × 234
= 468 feet

Solution: The parks department will need 468 feet of fencing.

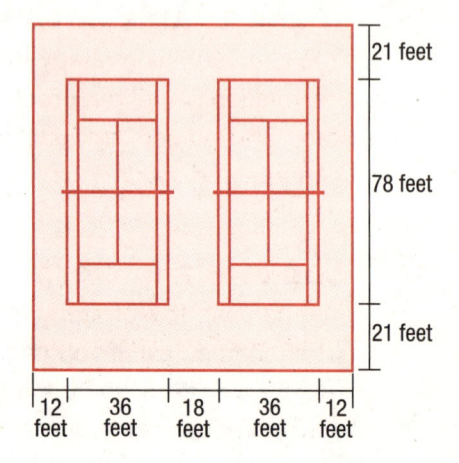

Look Back

You can use logical thinking to check your answer.

36 + 36 = 72 feet	*width of 2 courts*
12 + 18 + 12 = 42 feet	*space between and on sides*
72 + 42 = 114 feet	*total width*
21 + 21 = 42 feet	*space at ends of court*
42 + 78 = 120 feet	*total length*
114 + 120 + 114 + 120 = 468 feet	*perimeter* ✔

8

Lesson 1: **Draw a Diagram**

Problem Solving Practice
Find Out • Plan • Solve • Look Back

Try the *draw a diagram* strategy to solve these problems. Show your work.

1. Paula is hanging 5 posters on the wall along the length of a room. Each poster is $1\frac{1}{4}$ foot wide. There are 6 inches of space between posters. There is 1 foot of space before the first poster and after the fifth poster. How long is the room?

2. A hot dog vendor at a ballpark walks up and down the steps in his area during the ball game. He starts at row 1, walks up 9 rows, up 4 rows, up 6 rows, down 11 rows, up 7 rows, and down 3 rows. Then he runs out of hot dogs. He returns to row 1 to get more. How many rows does he walk down?

3. Mr. Suarez delivers milk to grocery stores. He drives 8 miles north from the plant to Bill's Market and then 5 miles west to Speedy Stop. From there, he drives 9 miles south to Frieda's Foods and then 5 miles east to Downtown Dave's. How far and in what direction does Mr. Suarez have to drive to get back to the plant?

4. At the town hall, there is a section of the yard that is 16 feet wide and 28 feet long. The mayor wants to construct a garden around a circular pond. What would be the area of the garden if she builds the largest pond possible?

5. The bumper cars are just inside the amusement park entrance. The haunted house is between the roller coaster and the Ferris wheel. The roller coaster is next to the bumper cars. Which attraction is farthest from the entrance to the park?

6. The community center tablecloth is 70 square feet. The dining table is 4 feet wide and 9 feet long. The tablecloth hangs over the ends of the table by 6 inches at each end. How much overhang is there on each side of the table?

NOTICE: Photocopying any part of this book is forbidden by law.

Part I: Strategies for Solving Problems

2 Make a Model

Sometimes it helps to use objects to model the solution to a problem. You can use the strategy *make a model or act out*.

Example

Dawn has 16 bricks to make a border for a square vegetable garden. Each plant needs 1 square foot of space. How many plants can she put in her garden?

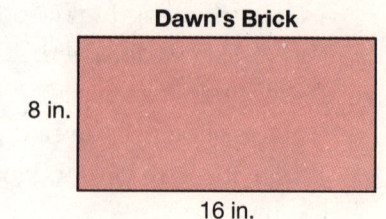

Dawn's Brick
8 in.
16 in.

Find Out

Think: *What facts do you know?*
- Dawn has 16 bricks, each 8 inches by 16 inches.
- Each plant needs 1 square foot of space.

Think: *What do you need to find out?*
You need to find out how many plants Dawn can put in her garden.

Plan

Think: *What strategy can you use?*
You can model the problem with paper "bricks."

Solve

First, cut out 16 grid paper bricks. Each "brick" should be 8 units wide by 16 units long.

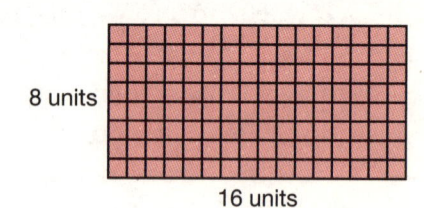

8 units
16 units

Arrange the bricks so that they enclose a square.

Next, find the inside length of one side of the square.

Change the units to feet.

56 inches = $4\frac{2}{3}$ feet

Then multiply to find the area.

$4\frac{2}{3} \times 4\frac{2}{3}$

$= \frac{14}{3} \times \frac{14}{3}$

$= \frac{196}{9} = 21\frac{7}{9}$ square feet

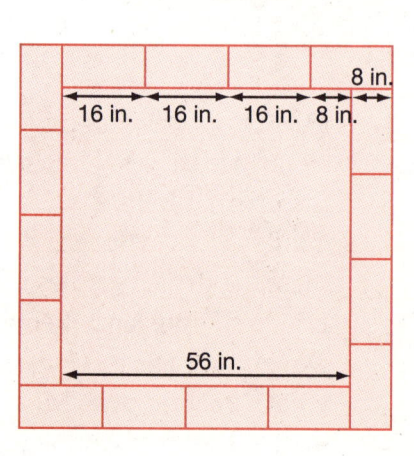

8 in.
16 in. 16 in. 16 in. 8 in.
56 in.

Solution: Dawn can put 21 vegetable plants in her garden.

Look Back

You can use estimation to check your answer.
56 inches are about 5 feet. $5 \times 5 = 25$ square feet
25 square feet is close to $21\frac{7}{9}$ square feet. ✔

Lesson 2: **Make a Model**

Problem Solving Practice
Find Out • Plan • Solve • Look Back

Try the *make a model* strategy to solve these problems. Show your work.

1. Mr. Wingate wants to build a square fenced-in play area. He has 20 sections of fencing. Each section is $3\frac{1}{2}$ feet long. What will be the area of the play area?

2. In a cake-eating contest, Todd ate $\frac{13}{16}$ of his cake. Ralph ate $\frac{5}{8}$ of his cake. Edward ate $\frac{11}{12}$ of his cake, and Oliver ate $\frac{5}{6}$ of his cake. Who ate the most cake?

3. A square sheet of paper is folded along the diagonal. It is folded in half again to bisect the largest angle. It is again folded in half to bisect the largest angle. Classify the resulting shape.

4. Gayle uses even numbers from 8 to 18 to label a number cube. The sum of opposite faces is always 26. Use the net for a cube below. Label each face with the correct number.

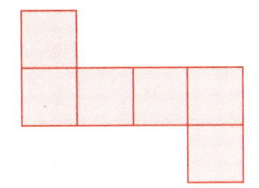

5. Start with the figure below. Reflect it across a horizontal line and then rotate it 90° counterclockwise around the dot. Now where is the triangle in relation to the square?

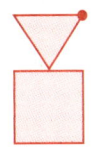

6. Simone has a regular 5-sided shape and 5 congruent acute isosceles triangles. What solid figure can she construct?

NOTICE: Photocopying any part of this book is forbidden by law.

Part I: Strategies for Solving Problems

3 Make an Organized List

You can use the strategy *make an organized list* when you need to find all the possible arrangements or outcomes for a problem.

Example

Ronnie is a cashier. She wants to use the fewest coins to give each customer change. What are the fewest coins she can use to give 27¢ as change to one customer and 33¢ as change to another customer?

Find Out

Think: *What facts do you know?*
Ronnie wants to use the fewest coins when she makes change.

Think: *What do you need to find out?*
What are the fewest coins needed to make 27¢ and 33¢?

Plan

Think: *What strategy can you use?*
You can make an organized list of all the combinations of coins for each amount.

Solve

Use the letters **q**, **d**, **n**, and **p** for quarter, dime, nickel, and penny. Make a list showing all the ways to make each amount of money. Circle the combinations that have the fewest coins.

27¢	
Coins	Number of Coins
27 p	27
22 p, 1 n	23
17 p, 1 d	18
17 p, 2 n	19
12 p, 1 d, 1 n	14
12 p, 3 n	15
7 p, 2 d	9
7 p, 1 d, 2 n	10
7 p, 4 n	11
2 p, 1 q	3
2 p, 2 d, 1 n	5
2 p, 1 d, 3 n	6
2 p, 5 n	7

33¢	
Coins	Number of Coins
33 p	33
28 p, 1 n	29
23 p, 1 d	24
23 p, 2 n	25
18 p, 1 d, 1 n	20
18 p, 3 n	21
13 p, 2 d	15
13 p, 1 d, 2 n	16
13 p, 4 n	17
8 p, 1 q	9
8 p, 2 d, 1 n	11
8 p, 1 d, 3 n	12
8 p, 5 n	13
3 p, 1 q, 1 n	5
3 p, 3 d	6
3 p, 2 d, 2 n	7
3 p, 1 d, 4 n	8
3 p, 6 n	9

Solution: Ronnie can use 3 coins to make 27¢ and 5 coins to make 33¢.

Look Back

You can make a model to check your answer. Use play money. ✔

Lesson 3: *Make an Organized List*

Problem Solving Practice
Find Out • Plan • Solve • Look Back

Try the *make an organized list* strategy to solve these problems. Show your work.

1. How many ways can you make 1 whole using fraction bars for $\frac{1}{2}$, $\frac{1}{4}$, and $\frac{1}{8}$?

2. Shana has a bag with 6 marbles. There is 1 marble of each color: red, yellow, blue, green, white, and purple. How many possible combinations are there if she randomly picks 2 marbles from the bag?

3. Members of the school chorus are having a bake sale to raise funds for a trip. They are taking orders for cakes to be delivered in two weeks.

   Cake Sale
   Design your own cake!

Cake	Filling	Icing
White	Apricot	Strawberry
Chocolate	Blueberry	Orange
Yellow	Peach	Vanilla
	Rasberry	Lemon

   How many different cakes are possible?

4. Dr. Uri bought 5 different kinds of plants for her office. She wants the tulips to be in the middle. How many arrangements are possible with tulips in the middle?

5. The Hansons are buying a car. The models are XT, XE, DL, LE, and LX. Each model comes in forest green, emerald green, midnight blue, and ruby red. The interior can be charcoal or tan. Mr. Hanson only wants a forest green or ruby red car. How many combinations can they choose from?

6. What is the greatest 3-digit number whose digits have a sum of 20? What is the greatest 3-digit multiple of 5 whose digits have a sum of 20?

Part I: Strategies for Solving Problems

4 Make a Table or Graph

You can organize information given in a problem is by using the strategy *make a table or graph*.

Example These are the numbers of CDs a singer sold during the first 10 weeks after it was released.

Week 1	85,000	Week 6	98,000
Week 2	97,000	Week 7	90,000
Week 3	103,000	Week 8	94,000
Week 4	101,000	Week 9	92,000
Week 5	105,000	Week 10	88,000

Between which two weeks did sales increase the most?

Find Out *Think: What facts do you know?*
You know how many CDs were sold each week.

Think: What do you need to find out?
You want to know between which two weeks sales increased the most.

Plan *Think: What strategy can you use?*
You can make a graph to find the changes in sales.

Solve Make a graph of the data.

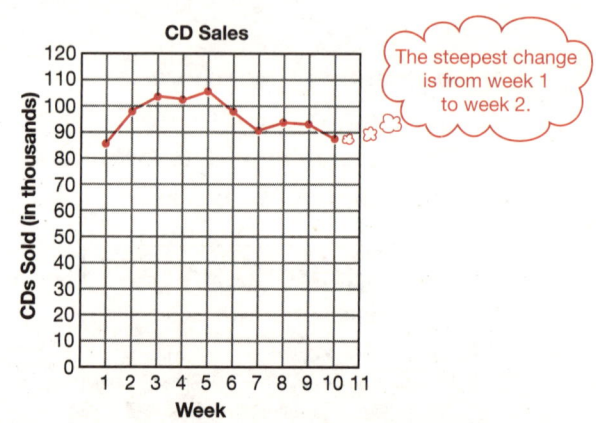

The steepest change is from week 1 to week 2.

Solution: Sales increased the most, by about 12,000, between week 1 and week 2.

Look Back You can make a table to check your answer.

CD Sales

Week	1	2	3	4	5	6	7	8	9	10
CDs Sold	85,000	97,000	103,000	101,000	105,000	98,000	90,000	94,000	92,000	88,000
Difference		12,000	6,000	−2,000	4,000	−7,000	−8,000	4,000	−2,000	−4,000

Sales increased the most between week 1 and week 2. ✔

Lesson 4: **Make a Table or Graph**

Problem Solving Practice
Find Out • Plan • Solve • Look Back

Try the *make a table or graph* strategy to solve these problems. Show your work.

1. Arnie recorded the low temperature in degrees Fahrenheit outside his window each day for a week in January.

Day	Temperature (Fahrenheit)
Monday	5°
Tuesday	2°
Wednesday	–3°
Thursday	0°
Friday	–5°
Saturday	1°
Sunday	–2°

   Between which 2 days was the temperature difference the greatest?

2. A sports company sells 3 different models of sneakers. These are last year's sales:

Fleet Feet	101,000 pairs
Snappy Sneaks	103,000 pairs
Rapid Runners	110,000 pairs

   The sales manager prepares a bar graph of this data. He uses a broken vertical scale that begins at 100,000 and goes to 110,000. Why would this graph convince the president to sell only Rapid Runners? Explain your thinking.

   _____
   _____
   _____

3. A sales representative is paid a 4% commission on the first $100,000 of sales. There is a 3% commission on the next $50,000 of sales and 2% on the next $50,000. How much does a sales representative earn for sales of $175,000?

4. One day last month, Suzanne had $128. On each subsequent day, she spent half of the money she had. On which day did she have less than $1 remaining?

5. The Family Fitness Center signed up 15 new members. The ages of the new members are given below.

   35, 48, 45, 52, 33, 40, 29, 55, 37, 42, 51, 33, 49, 30, 41

   Find the range, median, and mode of the ages.

Part I: Strategies for Solving Problems

5 Look for a Pattern

You can solve some problems by using the strategy *look for a pattern*.
Find the rule for a pattern and then use it to solve the problem.

Example   In certain bacteria, the number of cells doubles every 15 minutes. If you start with 1 cell, how many cells will there be after 3 hours?

Find Out   **Think:** *What facts do you know?*
You know that the number of cells doubles every 15 minutes.

Think: *What do you need to find out?*
You want to know how many cells there will be after 3 hours if you start with 1 cell.

Plan   **Think:** *What strategy can you use?*
You can look for a pattern and continue it for 3 hours.

Solve   Write a pattern to show that the number of cells doubles every 15 minutes. Since 15 minutes = $\frac{1}{4}$ hour, continue the pattern for 12 steps.

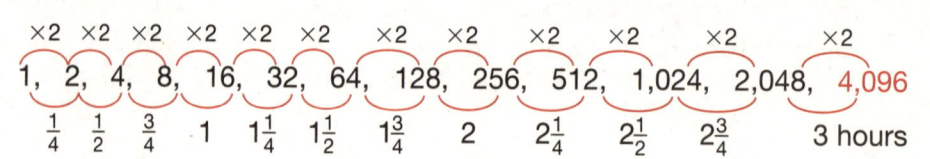

$$\begin{array}{cccccccccccccc}
& \times 2 & \times 2 & \times 2 & \times 2 & \times 2 & \times 2 & \times 2 & \times 2 & \times 2 & \times 2 & \times 2 & \times 2 \\
1, & 2, & 4, & 8, & 16, & 32, & 64, & 128, & 256, & 512, & 1{,}024, & 2{,}048, & 4{,}096 \\
& \frac{1}{4} & \frac{1}{2} & \frac{3}{4} & 1 & 1\frac{1}{4} & 1\frac{1}{2} & 1\frac{3}{4} & 2 & 2\frac{1}{4} & 2\frac{1}{2} & 2\frac{3}{4} & \text{3 hours}
\end{array}$$

Solution: There will be 4,096 cells after 3 hours.

Look Back   You can use multiplication to check your answer.

$\frac{1}{4}$ hour	$2 \times 1 = 2$
$\frac{1}{2}$ hour	$2 \times 2 = 4$
$\frac{3}{4}$ hour	$2 \times 4 = 8$
1 hour	$2 \times 8 = 16$
$1\frac{1}{4}$ hours	$2 \times 16 = 32$
$1\frac{1}{2}$ hours	$2 \times 32 = 64$
$1\frac{3}{4}$ hours	$2 \times 64 = 128$
2 hours	$2 \times 128 = 256$
$2\frac{1}{4}$ hours	$2 \times 256 = 512$
$2\frac{1}{2}$ hours	$2 \times 512 = 1{,}024$
$2\frac{3}{4}$ hours	$2 \times 1{,}024 = 2{,}048$
3 hours	$2 \times 2{,}048 = 4{,}096$ ✔

Lesson 5: **Look for a Pattern**

Problem Solving Practice
Find Out • Plan • Solve • Look Back

Try the *look for a pattern* strategy to solve these problems. Show your work.

1. Look at the pattern below.

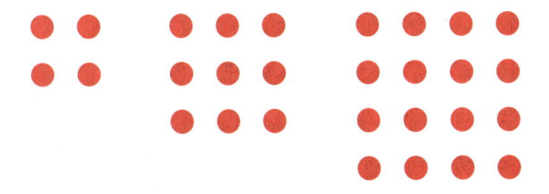

   If the pattern continues, how many dots will be in the eighth figure?

2. A school district bought new math books for grades 3 through 8. The third-grade book contains 360 pages. The books for the other grades contain 10% more pages, rounded to the nearest whole number, than the previous grade. How many pages are in the eighth-grade math book?

3. What are the next three numbers in this pattern? What is the rule for the pattern?

   $\frac{1}{3}, \frac{2}{9}, \frac{4}{27}, \underline{?}, \underline{?}, \underline{?}$

4. Uranium-238 has a half-life of about 4.5 billion years. That means that half of the amount of U-238 will decay in 4.5 billion years. How much of a 4-kilogram sample of U-238 will be left after 13.5 billion years?

5. Mrs. Arthur is creating a design for the border of a wall.

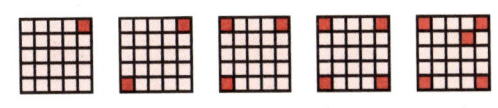

   If she continues the pattern, what will be the next three figures in the design?

6. If this pattern continues, what is the next shape in the pattern?

NOTICE: Photocopying any part of this book is forbidden by law.

Part I: Strategies for Solving Problems

6 Predict and Test

Sometimes you don't know how to solve a problem. You can try the *predict and test* strategy. Make a guess, and then test it. If your guess is not correct, you can use that guess to make a better guess.

Example A library has 875 films. There are 137 more videos than DVDs. How many DVDs are there?

Find Out **Think:** What facts do you know?
Of the 875 films, there are 137 more videos than DVDs.

Think: What do you need to find out?
You want to find out how many DVDs there are.

Plan **Think:** What strategy can you use?
You can use your number sense and then predict and test.

Solve Try 500 for DVDs.
500 + 137 = 637   *There would be 637 videos.*
637 + 500 = 1,137 *Add the videos and DVDs.*
                   Too high.

Try 300 for DVDs.
300 + 137 = 437   *There would be 437 videos.*
437 + 300 = 737   *Add the videos and DVDs.*
                   Too low.

Try 369 for DVDs.
369 + 137 = 506   *There would be 506 videos.*
506 + 369 = 875   *Add the videos and DVDs.*
                   This is correct.

Solution: There are 369 DVDs.

Look Back You can start with the answer and work backward to see if you get the information in the problem.

875 − 369 = 506   *There are 369 DVDs and 506 videos.*
506 − 369 = 137   *There are 137 more videos than DVDs.* ✔

Lesson 6: **Predict and Test**

Problem Solving Practice
Find Out • Plan • Solve • Look Back

Try the *predict and test* strategy to solve these problems. Show your work.

1. A park ranger estimates that there are 189 deer in the park. There are 63 more female deer than male deer. How many male deer are in the park?

2. Priscilla earned 20% more money than Patsy. Together, they earned $123.20. How much money did Priscilla earn?

3. A barrel holds 22 times more rainwater than a bucket. The bucket holds 52.5 fewer gallons of rainwater than the barrel. How much rainwater does the barrel hold?

4. An elephant and her calf weigh 9,000 pounds together. The mother weighs 8,500 pounds more than her calf. How much does the calf weigh?

5. Jeff drew triangle *ABC*. Angle *A* equals twice the measure of angle *B*. Angle *C* equals angle *A*. What are the measures of each of the 3 angles in triangle *ABC*?

6. The circle graph shows the voting for Student Council President. Christine got 6% more votes than Bob got. What percent of the votes did Christine get?

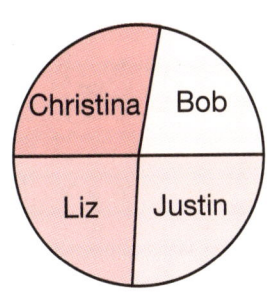

NOTICE: Photocopying any part of this book is forbidden by law.

Part I: Strategies for Solving Problems

7 Use Logical Thinking

Sometimes you need to think about how the information you know fits together. You can use the strategy *use logical thinking* to solve a problem.

Example The tallest player on the 2000 Olympics U.S. Women's Basketball Team was 6 feet 7 inches. She was 13 inches taller than the shortest player. Another player was 7 inches taller than the shortest player. What are the heights of the three players?

Find Out **Think:** *What facts do you know?*
You know the height of the tallest player and how that height compares with the heights of the other two players.

Think: *What do you need to find out?*
You want to find the heights of the three players.

Plan **Think:** *What strategy can you use?*
You can use logical thinking to find the heights of the players.

Solve Make a chart showing each player. Find the heights using the clues.

Tallest player	6 feet 7 inches	
Shortest player	6 feet 7 inches − 1 foot 1 inch 5 feet 6 inches	← 13 inches shorter than the tallest player
Another player	5 feet 6 inches + 7 inches 6 feet 1 inch	← 7 inches taller than the shortest player

Solution: The tallest player was 6 feet 7 inches.
The shortest player was 5 feet 6 inches.
Another player was 6 feet 1 inch.

Look Back You can work backward to check your answer.

The tallest player is 13 inches taller than shortest.
  5 feet 6 inches + 13 inches = 6 feet 7 inches ✓

The shortest player is 7 inches shorter than another player.
  6 feet 1 inch − 7 inches = 5 feet 6 inches ✓

Lesson 7: Use Logical Thinking

Problem Solving Practice
Find Out • Plan • Solve • Look Back

Try the *use logical thinking* strategy to solve these problems. Show your work.

1. Students are raising money to buy a new computer for the school library. Janet collected $37 more than Ken. Rob collected 3 times as much as Lucia. Janet collected $15 less than Rob. Ken collected $56. How much did Janet, Rob, and Lucia each collect?

2. Nikki's age is a prime number between 9 and 40. Joe is 6 years older. His age is a perfect square. How old is Nikki?

3. Iris, César, and Gary are in the school band. They play the violin, flute, and drums.
   - The violinist sits in the first row.
   - César does not play the drums.
   - Iris does not play the violin.
   - César doesn't sit in the first row.

   Which instrument does each student play?

4. Renie put paper covers on five books. Her Spanish book is between her History and Math books. Her English book is below her History book. Her Science book is at the top. Write the names on the books.

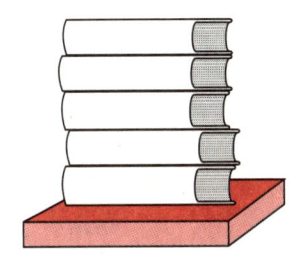

5. Four friends have after-school jobs. Anna earns $\frac{1}{3}$ more than Phyllis earns. Miguel earns $\frac{5}{6}$ of what Anna earns. Doug earns $\frac{4}{5}$ of what Miguel earns. Phyllis earns $9 per hour. How much do Anna, Miguel, and Doug earn?

6. Will's father was 26 years old when Will was born. Will's grandfather got married when he was 25 years old. Will's father was born 4 years later. How old was Will's grandfather when Will was born?

NOTICE: Photocopying any part of this book is forbidden by law.

Part I: Strategies for Solving Problems

8 Work Backward

Sometimes you know the amount at the end of a situation, but you need to find a missing part to solve the problem. You can use the strategy *work backward*. Start at the end and work back to the beginning.

Example

Julio spent $\frac{1}{3}$ of his money at a computer outlet. Then he spent $\frac{1}{2}$ of the remaining money at the sports store. After spending $11 for lunch, he came home with $4. How much money did Julio start with?

Find Out

Think: *What facts do you know?*
Julio spent $\frac{1}{3}$ of his money, $\frac{1}{2}$ of the remaining money, and then $11. He had $4 left.

Think: *What do you need to find out?*
You want to find out how much money Julio started with.

Plan

Think: *What strategy can you use?*
Since you know how much money Julio came home with and how much he spent, work backward.

Solve

$4 + $11 = $15   *He has $4 after spending $11 on lunch.*

$15 = $\frac{1}{2}$ × s   *$15 is $\frac{1}{2}$ of the money he had at the sports store.*
s = $30

$30 = $\frac{2}{3}$ × c   *He spent $\frac{1}{3}$ of his money at the computer outlet.*
c = $45   *So $45 is $1 - \frac{1}{3}$, or $\frac{2}{3}$, of the money he had.*

Solution: Julio started with $45.

Look Back

You can start with the answer and work forward to check.

$45 × $\frac{1}{3}$ = $15   *Spent $\frac{1}{3}$ of his money at a computer outlet.*
$45 − $15 = $30

$30 × $\frac{1}{2}$ = $15   *Spent $\frac{1}{2}$ of the remaining money at a sports store.*
$30 − $15 = $15

$15 − $11 = $4   *Spent $11 for lunch; had $4 left.* ✔

Lesson 8: **Work Backward**

Problem Solving Practice
Find Out • Plan • Solve • Look Back

Try the *work backward* strategy to solve these problems. Show your work.

1. Anita has an average of 89.6 on 5 math tests. Her scores are 85, 89, 90, and 92. What is her score on the fifth test?

2. Jorge fills a vase shaped like a rectangular prism with 2,600 cubic centimeters of water. The vase is 13 centimeters long and 12.5 centimeters wide. How high is the vase?

3. Alexis has $100 and a $25 gift certificate for the discount store. She plans to buy a new DVD player, 2 DVDs, and some CDs. How many CDs can Alexis buy?

   Super Saturday Sale
   DVD Players $60
   All DVDs—$17.50 each
   All CDs—$13.00 each

4. The Nolans are going to a movie at 2:30 P.M. The bus ride takes 25 minutes. It takes 10 minutes to walk from the bus stop to the theater. The Nolans want to allow $1\frac{1}{4}$ hours for lunch before the movie. What is the latest they should leave home?

5. Students at a high school are trying to raise $1,000 in pennies for charity. So far, the freshmen have 13,568 pennies. The sophomores have 24,997 pennies, and the juniors have 23,007 pennies. The seniors have 28,312 pennies. How many more pennies do they need?

6. In Victor's town, 45% of eligible voters voted in the last election. The town hopes to raise this to 75% by getting 900 more people to vote in the next election. How many eligible voters are there in the town?

NOTICE: Photocopying any part of this book is forbidden by law.

Part I: Strategies for Solving Problems

9 Solve a Simpler Problem

Sometimes the numbers in a problem are large or confusing. You can use the strategy *solve a simpler problem*. Change the numbers in the problem to smaller numbers that are easier to work with. Decide how to solve the problem. Then use your method to solve the original problem.

Example

A radio station had a call-in contest. People had to give the correct answer to a riddle to win a prize. Of the callers, 35 people gave the correct answer. They were 2% of the people who called in. How many people called in?

Find Out

Think: *What facts do you know?*
You know that 35 people called in the correct answer and they were 2% of the people who called in.

Think: *What do you need to find out?*
You want to find out how many people called in.

Plan

Think: *What strategy can you use?*
You can solve a simpler problem. Then use your method to solve the original problem.

Solve

Simpler Problem:
Assume that 35 people were 50% of all the people.

50% = 0.5
35 ÷ 0.5 = 70

Divide the number who gave the correct answer by the percent they represent.

Original Problem:
2% = 0.02
35 ÷ 0.02 = 1,750

Use the real numbers.

Solution: In all, 1,750 people called in.

Look Back

To check your answer, you can find 2% of 1,750.
1,750 × 0.02 = 35   *You started with 35.* ✔

Lesson 9: **Solve a Simpler Problem**

Problem Solving Practice
Find Out • Plan • Solve • Look Back

Try the *solve a simpler problem* strategy to solve these problems. Show your work.

1. Angie and Celia are looking at a checkerboard.

   Angie says there are 64 squares. Celia says there are more. Who is correct? How many squares are there? Explain your answer.

   _____
   _____
   _____
   _____

2. Dan has $3.70. He has a total of 24 quarters, nickels, and dimes. How many of each coin does Dan have?

3. Erika and her family are driving to visit relatives in Canada. They have already driven 348 miles, which is 30% of the distance. How many more miles do they still have to drive?

4. All people have tiny mites that live on their bodies. Eyelash mites are 0.4 millimeter long. How many times longer than an eyelash mite is a string that is 1 meter long?

5. A company manufactures drinking glasses. The workers pack four glasses to a box. They can pack 12 boxes in a carton. How many cartons will they need to pack 6,576 glasses?

6. The ruby-throated hummingbird migrates from the United States to the Yucatán Peninsula in Mexico. It flies 500 miles across the Gulf of Mexico nonstop at a speed of 30 miles each hour. The hummingbird makes about 60 wing beats every second. How many wing beats does the hummingbird make over the Gulf of Mexico?

NOTICE: Photocopying any part of this book is forbidden by law.

Part I: Strategies for Solving Problems

10 Write a Formula or an Equation

Sometimes you need to find a missing amount to solve a problem. You can use the *write a formula or an equation* strategy.

Example The Great Pyramid of Cheops in Egypt has a rectangular base that is about 755 feet wide and 620 feet long. It is about 450 feet tall. About what is the volume of the pyramid?

Find Out *Think:* What facts do you know?
You know that the Great Pyramid has a rectangular base.
length = 620 feet, width = 755 feet, height = 450 feet

Think: What do you need to find out?
You want to know the approximate volume of the pyramid.

Plan *Think:* What strategy can you use?
You can write an equation to find the volume.

Solve Use the formula for the volume of a pyramid.

Volume = $\frac{1}{3}$ × area of base × height
Volume = $\frac{1}{3}$ × length × width × height

Write the equation using the measurements you know.

Volume = $\frac{1}{3}$ × length × width × height
Volume = $\frac{1}{3}$ × 620 × 755 × 450
Volume = 70,215,000 cubic feet

Solution: The volume of the pyramid is about 70 million cubic feet.

Look Back You can estimate to check the reasonableness of your answer.

Volume = $\frac{1}{3}$ × length × width × height
Volume ≈ $\frac{1}{3}$ × 600 × 800 × 500
Volume ≈ 80,000,000 cubic feet
80,000,000 cubic feet is close to 70,215,000 cubic feet. ✓

Lesson 10: *Write a Formula or an Equation*

Problem Solving Practice
Find Out • Plan • Solve • Look Back

Try the *write an equation* strategy to solve these problems. Show your work.

1. The weight of an object on the moon is $\frac{1}{6}$ the object's weight on Earth. Suppose a fully equipped astronaut weighs 300 pounds on Earth. How much would the astronaut weigh on the moon?

2. In 1864, Abraham Lincoln received 2,218,388 votes. This was about 55% of all the votes cast. How many people voted in the 1864 presidential election?

3. Hair grows about $\frac{1}{2}$ inch every month. Nails grow about $\frac{1}{8}$ inch every month. How much faster does hair grow than nails in a month?

4. A hospital has collected $833,333 in donations. This is $\frac{5}{9}$ of the total they hope to raise. How much money do they hope to raise? Round your answer to the nearest $100,000.

5. Henry buys a used car for $4,500. How much does he pay in all if the sales tax is 6%?

6. Ashley takes the Acme Taxi Service home from the airport. These are its rates:

> **ACME Taxi Service**
> $2.50 for the first mile
> $1.35 for each additional mile
> $6.50 extra for airport service

Ashley's trip is 17 miles. How much does the ride cost?

NOTICE: Photocopying any part of this book is forbidden by law.

Part II: Applying the Strategies—Solving Problems with Rational Numbers

11 Operations with Whole Numbers, Fractions, and Decimals

Whole numbers, fractions, and decimals can be used to solve problems that involve computation, estimation, and comparison.

Example 1

This chart shows the total amount Jerry earned from his lawn mowing business for the first five weeks. If the pattern continues, what will be the total amount of his earnings after the eighth week?

Jerry's Lawn-mowing Earnings

Week	Total Amount Earned After Each Week
1	$32
2	$48
3	$72
4	$108
5	$162

Find Out

The chart shows the total amount he earned after weeks 1, 2, 3, 4, and 5. You want to find the total amount he will have earned after week 8.

Plan

You can use the *look for a pattern* strategy to solve this problem. Look for a relationship between consecutive numbers.

Solve

Each week the total is more than the week before, so try dividing to find a pattern.

Week 2 ÷ Week 1 = $48 ÷ $32 = 1.5
Week 3 ÷ Week 2 = $72 ÷ $48 = 1.5
Week 4 ÷ Week 3 = $108 ÷ $72 = 1.5
Week 5 ÷ Week 4 = $162 ÷ $108 = 1.5

So, each week's total is 1.5 times the previous week.

Multiply the total amount for each week after week 5 by 1.5.
Week 6 $162 × 1.5 = $243
Week 7 $243 × 1.5 = $364.50
Week 8 $364.50 × 1.5 = $546.75

Solution: After the eighth week, Jerry's total earnings will be $546.75.

Look Back

You can work backward to check your answer.
Week 8 = $546.75
Week 7 = $546.75 ÷ 1.5 = $364.50
Week 6 = $364.50 ÷ 1.5 = $243
Week 5 = $243 ÷ 1.5 = $162

Week 4 = $162 ÷ 1.5 = $108
Week 3 = $108 ÷ 1.5 = $72
Week 2 = $72 ÷ 1.5 = $48
Week 1 = $48 ÷ 1.5 = $32 ✓

Lesson 11: **Operations with Whole Numbers, Fractions, and Decimals**

Try It

Example 2

Adam, Beth, Carly, and Della held their first meeting as elected class officers. Each one shook hands with each of the other three. What fraction of the handshakes was between a boy and a girl?

Find Out You know 1 boy and 3 girls shook hands with each other. You want to find out what fraction of the handshakes was between a boy and a girl.

Plan You can use the *draw a diagram* strategy to help you see the solution to the problem.

Solve Draw Adam's handshakes.

Adam — Beth, Carly, Della

Draw Beth's handshakes.
Cross out any repeats.

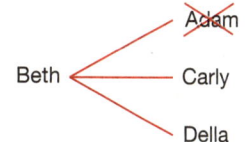

Beth — ~~Adam~~, Carly, Della

Draw Carly's handshakes.
Cross out any repeats.

_____ — _____, _____, _____

Draw Della's handshakes.
Cross out any repeats.

_____ — _____, _____, _____

How many total handshakes were there? _____

How many handshakes were between Adam and a girl? _____

What fraction of the total handshakes was between Adam and a girl? _____

Solution: _____ of the handshakes were between a boy and a girl.

Look Back You can also solve this problem by making an organized list.
List all the pairs of handshakes. Cross off any repeats.

_____

_____

Write a fraction. $\frac{\text{Adam's handshakes}}{\text{Total handshakes}}$ = _____

Is the answer the same? _____

NOTICE: Photocopying any part of this book is forbidden by law.

Practice Test Questions
Find Out • Plan • Solve • Look Back

1. At a baseball park, $\frac{5}{6}$ of the seats were filled. The baseball park has 18,000 seats. How many seats were filled? (Hint: Try the *look for a pattern* strategy.)

   A 9,000
   B 12,000
   C 15,000
   D 18,00

2. The balcony in the school auditorium has 8 rows with 12 seats each and 8 rows with 9 seats each. Which expression can you use to find the total number of seats in the balcony? (Hint: Try the *draw a diagram* strategy.)

   A $(8 + 12) \times (8 + 9)$
   B $8 \times (12 + 9)$
   C $12 \times (9 + 8)$
   D $8 \times 12 \times 9$

3. Jessica simplified an expression using the order of operations and got an answer of 4. If Jessica's answer is correct, which expression did she simplify? (Hint: Try the *predict and test* strategy.)

   A $4 \times 4 - 4 \div 4$
   B $3 \times 6 - (12 \div 3)$
   C $8 - 4 \times 2 + 5$
   D $(3^2 + 5) \div 7 \times 2$

4. A fifth-grade class raised $100 in a walk-a-thon. A seventh-grade class raised 10 times as much. The whole school raised 10 times as much as the seventh-grade class. How much money did the whole school raise?

   A $120   C $10,000
   B $1,000   D $100,000

5. The Carson family drove at an average speed of 52 miles an hour for $2\frac{3}{4}$ hours. What distance did they drive?

   A 143 miles   C $54\frac{3}{4}$ miles
   B 104 miles   D Not here

6. Jeremy bought this DVD player.

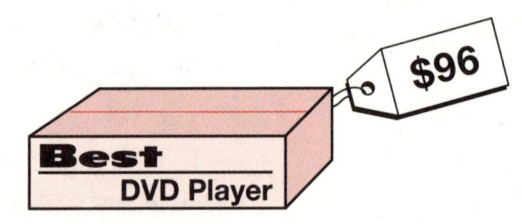

   Cindy paid $\frac{7}{8}$ as much for her DVD player. How many more dollars did Jeremy pay for his DVD player?

   Record your answer. Then fill in the bubbles. Be sure to use the correct place value.

Lesson 11: Operations with Whole Numbers, Fractions, and Decimals

7. There were 56 signers of the Declaration of Independence. Of these, $\frac{1}{8}$ were born in Virginia and $\frac{1}{14}$ were born in South Carolina. How many more signers were from Virginia? (Hint: Try the *solve a simpler problem* strategy.)

8. Every student in Ms. Diaz's class is in the theater club or the camera club.

Club Membership

Theater Club $\frac{3}{5}$ — $\frac{1}{10}$ — Camera Club ?

Three fifths of the students are only in the Theater Club. One tenth are in both the Theater Club and the Camera Club. What fraction of the students is only in the Camera Club? (Hint: Try the *use logical thinking* strategy.)

9. One week, the most-watched TV show had a rating of 12.4. A ratings point represents 1.07 million homes. The second most-watched program had a rating of 12.0. How many more homes watched the most-watched TV show? Show your work. (Hint: Try the *write an equation* strategy.)

10. Tonia has 65 cents in her wallet. She has at least 1 quarter, 1 dime, and 1 nickel. If she has 7 coins in her wallet, what coins does she have?

11. Andy needs a new sweater. Which sweater will cost less? Explain.

CREW NECK Sweaters	V-NECK Sweaters
Originally $30	Originally $32
Now $\frac{2}{3}$ off!!!	Now $\frac{3}{4}$ off!!!

_____
_____
_____

12. This table shows the population of Florida for 6 consecutive census years.

Population of Florida

Census Year	Population
1950	2,771,305
1960	4,951,560
1970	6,791,418
1980	9,746,961
1990	12,937,926
2000	15,982,378

Between which two census years did the population of Florida show the **least** increase?

Strategies: Diagram • Model • Organized List • Table/Graph • Pattern • Predict & Test
Logical Thinking • Work Backward • Simpler Problem • Formula or Equation

Part II: Applying the Strategies—Solving Problems with Rational Numbers

12 Operations with Integers

Integers are the counting numbers, their opposites, and zero. Operations with integers are useful in solving a variety of real-world problems that involve temperature, altitude, and money.

Example 1

Paul recorded the temperature at 10:00 A.M. for four days. On Tuesday, the temperature was 2 degrees warmer than on Monday. On Wednesday, the temperature was 4 degrees cooler than on Tuesday. On Thursday, the temperature was 5 degrees warmer than on Wednesday. The temperature was 3° Celsius on Thursday. What was the temperature on Monday?

Find Out

You want to find the temperature on Monday. You know the temperature on Thursday was 3° Celsius. You know how the temperature changed from day to day.

Plan

You can use the *work backward* strategy to solve this problem.

Solve

Start with Thursday's temperature.

The temperature was warmer on Thursday than on Wednesday.

3°C − 5°C = −2°C   *Subtract to find Wednesday's temperature.*

The temperature was cooler on Wednesday than on Tuesday.

−2°C + 4°C = 2°C   *Add to find Tuesday's temperature.*

The temperature was warmer on Tuesday than on Monday.

2°C − 2°C = 0°C   *Subtract to find Monday's temperature.*

Solution: The temperature on Monday was 0°C.

Look Back

How can you check your answer? Start with Monday's temperature. Use the information in the problem and work forward to see if you get Thursday's temperature.

0 + 2 − 4 + 5 = 3 ✔

32   NOTICE: Photocopying any part of this book is forbidden by law.

Lesson 12: **Operations with Integers**

Try It

Example 2 A diver descends 8 feet below the surface of the water to observe a school of fish. Next, she descends to a depth that is 4 times deeper to study the bottom of a coral formation. Then she ascends 20 feet to study the top of the coral formation. How far below the surface of the water was the diver at the top of the coral formation?

Find Out You need to find how far below the surface the diver is after she ascends the 20 feet.

Plan You can use the *draw a diagram* strategy. Draw a line to represent the surface of the water and use arrows to show the diver's movement.

Solve Draw a diagram. Label the parts.

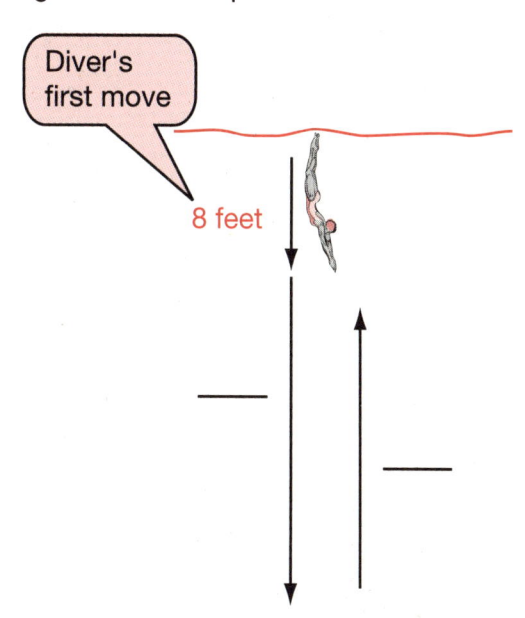

Diver's first move

8 feet

How will you label the diver's second move? Explain.

_____

How will you label her third move? Explain.

_____

How far is it from the tip of the last arrow to the water line?

_____

Solution: The diver is _____ below the surface of the water.

Look Back You can use operations with integers to check your work.

4 × −8 + 20 = _____   *Use the order of operations.*

33

Mathematics Problem Solving Coach, Level G: Strategies and Applications

Practice Test Questions
Find Out • Plan • Solve • Look Back

1. Bill wrote a number on a card. On the other side of the card, he wrote these clues to find his number.

   - Multiply it by 2.
   - Add 3.
   - The result is −5.

   What is Bill's number? (Hint: Try the *work backward* strategy.)

   A −5
   B −4
   C 2
   D 3

2. At the end of a game, Anya had a score of 6 and Brian had a score of −10. Clarence had a score that was halfway between Anya's and Brian's scores. What was Clarence's score? (Hint: Try the *draw a diagram* strategy.)

   A −2   C 2
   B 0   D 4

3. Natalie sells bracelets to her friends. Her profits and losses for 4 weeks were $24, −$3, $16, and −$5. What was her profit or loss for the month? (Hint: Try the *solve a simpler problem* strategy.)

   A −$32   C $32
   B $0   D $48

4. Laurie is charting the temperature in degrees Fahrenheit every hour. If this pattern continues, what will be the next temperature?

   17°F, 8°F, −1°F, −10°F

   A 19°F
   B 10°F
   C −1°F
   D Not here

5. The elevation of New Orleans is −8 feet. The elevation of Washington, D.C., is 1 foot. What is the difference in their elevations in feet?

   Record your answer. Then fill in the bubbles. Be sure to use the correct place value.

Lesson 12: **Operations with Integers**

6. A 177-pound boxer is training for the Olympics. He wants to qualify as a middleweight boxer with a weight of 165 pounds. He figures that he can safely lose 2 pounds per month. How many months will it take him to reach middleweight? Explain your answer. (Hint: Try the *write an equation* strategy.)

_____
_____
_____
_____

7. Marcia earns $6 per hour at her yard-cleaning business. Last weekend, she paid her brother $12 to help her. After she paid her brother, she had $66. How many hours did she work? Explain the strategy you used. (Hint: Try the *work backward* strategy.)

_____
_____
_____
_____
_____

8. Nestor created this number pattern to challenge his friend Larry. What is the next number in the pattern? (Hint: Try the *look for a pattern* strategy.)

$$-48, 24, -12, 6, \underline{}$$

9. In this square, the sum of the numbers along each row, column, and diagonal is the same. What is the missing number in this square?

?	-4	3
2	0	-2
-3	4	-1

10. Roger wrote these values for powers of -1.

$$(-1)^2 = (-1)(-1) = 1$$
$$(-1)^3 = (-1)(-1)(-1) = -1$$
$$(-1)^4 = (-1)(-1)(-1)(-1) = 1$$
$$(-1)^5 = (-1)(-1)(-1)(-1)(-1) = -1$$

What will Roger write for the value of $(-1)^9$? Explain your answer.

_____
_____
_____
_____

11. Julia played a game with this spinner. What is the lowest total she could get in 8 spins?

Strategies: Diagram • Model • Organized List • Table/Graph • Pattern • Predict & Test
Logical Thinking • Work Backward • Simpler Problem • Formula or Equation

Part II: Applying the Strategies—Solving Problems with Rational Numbers

13 Operations with Rational Numbers

The set of *rational numbers* includes whole numbers, fractions, decimals that terminate or repeat, and integers.

Example 1

Danielle has an emergency money jar that she puts money in when she has it and takes money out of when she needs it. Last Sunday, she had $100.00 in the jar. She took out $23.75 on Monday and put back $15.25 on Tuesday. On Thursday, she took out $17.50, and on Saturday, she put back $10.00. How much is in the jar now?

Find Out You need to find how much money is in the jar now. You know how much was in the jar to start with and how many withdrawals and deposits Danielle made.

Plan You can use the *solve a simpler problem* strategy. Break the problem into simpler parts and solve each part.

Solve Add up her withdrawals.
−$23.75 + −$17.50 = −$41.25   *A negative + a negative = a negative.*

Add up her deposits.
$15.25 + $10.00 = $25.25   *A positive + a positive = a positive.*

Find the total of her withdrawals and deposits.
−$41.25 + $25.25 = −$16.00

Add the total to $100.00.
$100.00 + −$16.00 = $84.00

Solution: There is $84.00 in the jar now.

Look Back You can check your answer by starting with $84.00 and working backward.

$84.00 − $10.00 + $17.50 − $15.25 + $23.75 = $100.00 ✔

Lesson 13: **Operations with Rational Numbers**

Try It

Example 2

Jake, Keith, and Lori each drew a rational number card from a deck of number cards. The numbers on the cards were:

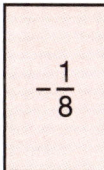

| $-\frac{1}{8}$ | 0.5 | $\frac{3}{4}$ |

Jake's number was less than Lori's number. The sum of Jake's number and Keith's number was less than 1. Keith's number was less than Jake's number. Which number did each person draw?

Find Out

You have three clues:
- Jake's number was less than Lori's.
- The sum of Jake's and Keith's numbers was less than 1.
- Keith's number was less than Jake's.

You want to find the number drawn by each person.

Plan

You can use the *use logical thinking* strategy to find the solution.

Think: *How can you make the numbers easier to compare?*

_____

Solve

Make a table. Draw an X to show a match between a person and a number.

	$-\frac{1}{8}$	0.5	$\frac{3}{4}$
Jake			
Keith			
Lori			X

The first and third clues tell you that Lori had the greatest number. Put an X in the box where Lori and $\frac{3}{4}$ meet.

What is Jake's number? Put an X in the table.

Which number, when added to Jake's, gives a sum less than 1?

_____

Solution: Jake drew _____, Keith drew _____, and Lori drew _____.

Look Back

What other problem-solving strategy can you use to check your solution?

_____

Mathematics Problem Solving Coach, Level G: Strategies and Applications

Practice Test Questions
Find Out • Plan • Solve • Look Back

1. Eldon recorded these withdrawals and deposits from his savings account:

Monday:	−$18.30
	+$39.00
Thursday:	+$49.00
Friday:	−$52.22

   Which is the total of his withdrawals and deposits? (Hint: Try the *solve a simpler problem* strategy.)

   A   −$88.52   C $17.48
   B   −$17.48   D $88.52

2. On Monday, $\frac{1}{12}$ of the 480 students in a school were absent. Which of the following questions could **not** be answered with the information provided? (Hint: Try the *use logical thinking* strategy.)

   A   How many students were on a field trip?
   B   How many students were present?
   C   What fraction of students was present?
   D   How many students were absent?

3. Kelly Ann found these lengths of wood in her cellar: $5\frac{3}{4}$ feet, 5.3 feet, $3\frac{1}{4}$ feet, 4.9 feet. What is the best estimate of the sum of the lengths? (Hint: Try the *solve a simpler problem* strategy.)

   A   15 feet   C   19 feet
   B   17 feet   D   21 feet

4. At a school bake sale, sixth graders bought $\frac{1}{5}$ of the muffins, and seventh graders bought $\frac{1}{3}$ of the muffins. Eighth graders bought the remaining 70 muffins. How many muffins were sold in all?

   A   70
   B   80
   C   150
   D   210

5. Melinda wrote an expression whose value was −9. Which expression did she write?

   A 4.5×2
   B $-\frac{3}{4} \times 12$
   C $-27 \times -\frac{1}{3}$
   D $12 \times -\frac{2}{3}$

6. One fourth of 196 seventh graders attended the football rally. How many seventh graders attended the rally?

   Record your answer. Then fill in the bubbles. Be sure to use the correct place value.

38

Lesson 13: **Operations with Rational Numbers**

7. On Monday, Collin deposited $25.50 in his checking account. On Tuesday, he deposited twice as much as he did on Monday. Then he wrote a check for $19.95. His balance was $166.75. How much money was in Collin's checking account before his $25.50 deposit? (Hint: Try the *work backward* strategy.)

8. Max has $120 in his account. If the interest rate is 0.02, how much interest will Max earn for $\frac{1}{2}$ year? (Hint: Try the *write an equation* strategy.)

$$\text{Interest} = \text{principal} \times \text{rate} \times \text{time}$$

9. What fraction of the students participates in all three activities? Explain. (Hint: Try the *use logical thinking* strategy.)

Student Activities

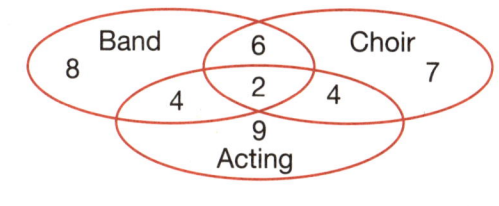

Band 8, Choir 7, Acting 9, Band∩Choir 6, Band∩Acting 4, Choir∩Acting 4, all three 2

10. BJ surveyed all of his classmates about their favorite pets. These are his results.

Favorite Pets

Dog	$\frac{1}{2}$ of students
Cat	$\frac{1}{4}$ of students
Fish	$\frac{3}{16}$ of students
Other	$\frac{1}{16}$ of students

Draw a graph that best shows how the fractional parts compare to one another and to the whole group. Explain your choice of graph.

11. Lisa and Patti are going to the movies 1 mile east of their home. When they had walked $\frac{1}{2}$ mile, they walked $\frac{3}{4}$ of the way back home to see if their friend Gina would join them. How far is it from Gina's house to the movies?

Strategies: Diagram • Model • Organized List • Table/Graph • Pattern • Predict & Test
Logical Thinking • Work Backward • Simpler Problem • Formula or Equation

Part II: Applying the Strategies—Solving Problems with Rational Numbers

14 Number Theory

Number theory is a branch of mathematics that includes the topics of factors, multiples, prime and composite numbers, and divisibility. A *prime number* has exactly two factors, 1 and itself. A *composite number* is a number that has more than two factors.

Example 1

For its grand opening, a supermarket offered these prizes.

| $2 gift certificate for every sixth customer | Discount coupon for every twelfth customer | $5 discount for every fifteenth customer |

Which customer will be the first to get all three prizes?

Find Out You know that every sixth, twelfth, and fifteenth customer will get a prize. You want to find out which customer will be the first to get all three prizes.

Plan You can use a *make a table* strategy to solve this problem.

Solve Write the multiples of 6, 12, and 15 in a table. The first multiple the three numbers have in common, the least common multiple, will be the customer who gets all three prizes.

6	12	18	24	30	36	42	48	54	60
12	24	36	48	60					
15	30	45	60						

Solution: The sixtieth customer will get all three prizes.

Look Back You can check your work by finding the least common multiple using the prime factorization method.

$6 = 2 \times 3$ $12 = 2 \times 2 \times 3$ $15 = 3 \times 5$

The least common multiple must have 2, 2, 3, and 5 for factors.
$2 \times 2 \times 3 \times 5 = 60$ ✔

Lesson 14: **Number Theory**

Try It

Example 2

There are 24 students in the marching band. They march in rectangular formations with the same number of students in each row and the same number of students in each column. Each row has at least 2 marchers. What are the possible formations?

Find Out

You know that there are 24 marchers. You know how the marchers line up. You need to find all the possible formations.

Plan

You can use the *make a model* strategy to solve this problem.

Solve

Use a counter to represent each marcher. This model represents one possible formation of 12 rows with 2 marchers each.

● ● ● ● ● ● ● ● ● ● ● ●
● ● ● ● ● ● ● ● ● ● ● ●

Can you form a rectangle with 3 marchers in a row? If so, how many rows would there be? _____

Can you form a rectangle with 5 marchers in a row? Why or why not?

_____

What other formations can you model?

_____
_____

Solution: The possible formations are:

12 rows with 2 marchers each,

_____ rows with _____ marchers each,

_____ rows with _____ marchers each,

_____ rows with _____ marchers each,

_____ rows with _____ marchers each, and

_____ rows with _____ marchers each.

Look Back

You can check your work by finding all the pairs of factors of 24, except 1 and 24.

12 × 2 = 24; ___ × 3 = 24; ___ × 4 = 24;

___ × ___ = 24; ___ × ___ = 24; ___ × ___ = 24 ✔

NOTICE: Photocopying any part of this book is forbidden by law.

41

Mathematics Problem Solving Coach, Level G: Strategies and Applications

Practice Test Questions
Find Out • Plan • Solve • Look Back

1. Sam, Tina, and Vin are all at the gym today. Sam goes every 2 days, Tina goes every 3 days, and Vin goes every 7 days. Which question can be answered with the information provided? (Hint: Try the *make a table* strategy.)

    A In how many days will they all be at the gym again?

    B How many more hours does Sam work out than Tina?

    C Who has been a member of the gym for the longest time?

    D Who is in the best physical condition?

2. Tim wants to arrange 36 books in equal piles that have at least 2 books. In how many ways can he do this? (Hint: Try the *make a model* strategy.)

    A 3   C 7
    B 4   D 9

3. Jenny found the prime factorization of the numbers in this pattern: 4, 8, 16, and 32. If the pattern continues, what will be the prime factorization of the next two numbers? (Hint: Try the *look for a pattern* strategy.)

    A 2^5 and 2^6

    B 2^5 and 2^7

    C 2^6 and 2^7

    D 2^6 and 2^8

4. Which two number cards show numbers that are both odd and prime?

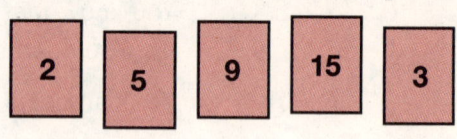

    [2] [5] [9] [15] [3]

    A 2 and 3
    B 3 and 5
    C 5 and 9
    D 9 and 15

5. Millie wrote this formula for generating even numbers:
    $2n - 2$ = an even number.
    If you substitute 1, 3, 5 and 7 in her formula, what even numbers will you get?

    A 0, 2, 4, 6
    B 2, 4, 6, 8
    C 4, 8, 12, 16
    D 0, 4, 8, 12

6. A number is divisible by both 3 and 5. What other divisor will the number have (other than 1)?

    Record your answer. Then fill in the bubbles. Be sure to use the correct place value.

42

Lesson 14: **Number Theory**

7. Whitney drew this factor tree, but she left out some of the numbers. What number is at the top of her factor tree? (Hint: Try the *work backward* strategy.)

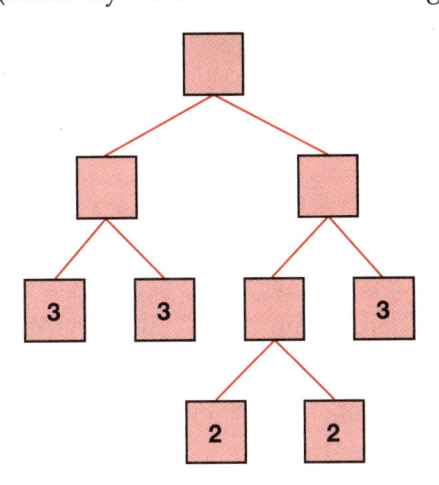

```
       [ ]
      /   \
   [ ]     [ ]
   / \     / \
  3   3  [ ]  3
         / \
        2   2
```

8. In Terry's game club, all the members can participate in 6-player games at the same time. All can participate in 8-player games at the same time. The number of club members is a number between 70 and 80. How many members are there? (Hint: Try the *use logical thinking* strategy.)

9. Two numbers have a greatest common factor of 24. Both of the numbers are greater than 70 and less than 100. Their difference is 24. What are the numbers? (Hint: Try the *predict and test* strategy.)

10. Twenty-four boys and 30 girls are sitting in separate rows in an auditorium. Each row has the same number of students. What is the greatest number of students that can be in a row? Explain your answer.

_____
_____
_____

11. There will be 240 people attending the school awards banquet.

> **Banquet Tables for Rent**
> Tables Available for
> 4, 6, 9, or 10 people

Which size tables can be rented so that every table will be filled? Explain.

_____
_____
_____

12. The buses to New City and to Glenview both leave the station at 7:00 A.M. New City buses leave the station every 45 minutes, and Glenview buses leave every 20 minutes. What is the next time that buses to the two cities will leave the station at the same time?

Strategies: Diagram • Model • Organized List • Table/Graph • Pattern • Predict & Test
Logical Thinking • Work Backward • Simpler Problem • Formula or Equation

Part II: Applying the Strategies—Solving Problems with Rational Numbers

15 Ratios, Rates, and Proportions

You use *ratios* to compare quantities having the same units. A *rate* is a ratio that compares quantities with different units. You can use a *proportion* to solve problems that compare ratios and rates.

Example 1

It takes Lauren 2.5 hours to make 12 greeting cards on her computer. At that rate, how many greeting cards can she make in 15 hours?

Find Out You know that Lauren can make 12 greeting cards in 2.5 hours. You need to find how many greeting cards she can make in 15 hours.

Plan You can use the *write an equation* strategy to solve this problem.

Solve Write a rate for the information you know.

$$\frac{12 \text{ greeting cards}}{2.5 \text{ hours}}$$

Let x represent the number of greeting cards in 15 hours. Write this as a rate.

$$\frac{x \text{ greeting cards}}{15 \text{ hours}}$$

Set the rates equal to form a proportion.

$\frac{12}{2.5} = \frac{x}{15}$

$2.5x = 12 \times 15$ *Write the cross products.*

$2.5x = 180$ *Multiply.*

$x = \frac{180}{2.5}$ *Divide to solve.*

$x = 72$

Solution: Lauren can make 72 greeting cards in 15 hours.

Look Back Substitute 15 for x in the proportion.

$\frac{12}{2.5} = \frac{72}{15}$ ✓

Lesson 15: **Ratios, Rates, and Proportions**

Try It

Example 2

In a school, the ratio of the number of seventh-grade boys to girls is 3 to 4. There are 6 more girls than boys. How many seventh graders are in the school?

Find Out

You know:
- The ratio of boys to girls is 3 to 4.
- There are 6 more girls than boys.

You need to find the total number of boys and girls.

Plan

You can use the *predict and test* strategy to solve this problem.

Think: *What fraction can you write for the ratio of boys to girls?* _____

Solve

Write ratios equivalent to the ratio of boys to girls and test to see if the denominator is 6 more than the numerator.

Ratio	Equivalent Ratio	Denominator − Numerator
$\frac{3}{4}$	$\frac{6}{8}$	$8 - 6 = 2$
$\frac{3}{4}$	$\frac{9}{12}$	$12 - 9 =$ _____
$\frac{3}{4}$	$\frac{15}{20}$	$20 -$ _____ $=$ _____
$\frac{3}{4}$	$\frac{18}{24}$	_____ $-$ _____ $=$ _____

There are _____ boys and _____ girls in the seventh grade.

Solution: There are _____ seventh graders in the school.

Look Back

What other strategy can you use to solve this problem?

_____

Use your strategy. Is your answer the same?

_____

NOTICE: Photocopying any part of this book is forbidden by law.

Mathematics Problem Solving Coach, Level G: Strategies and Applications

Practice Test Questions
Find Out • Plan • Solve • Look Back

1. A recipe for a batch of 20 cookies calls for $\frac{1}{4}$ cup of butter. How much butter would you need to make a batch of 50 cookies? (Hint: Try the *write an equation* strategy.)

   A $\frac{3}{4}$ cup

   B $\frac{5}{8}$ cup

   C $\frac{1}{2}$ cup

   D $\frac{3}{8}$ cup

2. You have a sheet of paper that measures 8 inches by 11 inches. You want to make a scale drawing of a room that measures 12 feet by 16 feet. Which scale would give you the largest drawing that would fit on the sheet? (Hint: Try the *predict and test* strategy.)

   A $\frac{1}{4}$ inch:1 foot

   B $\frac{1}{2}$ inch:1 foot

   C 1 inch:1 foot

   D 2 inches:1 foot

3. Two towns that are 25 miles apart are shown as 10 inches apart on a map. What is the scale of the map? (Hint: Try the *work backward* strategy.)

   A $\frac{1}{4}$ inch:1 mile

   B $\frac{1}{2}$ inch:1 mile

   C 0.4 inch:1 mile

   D 2.5 inches:1 mile

4. Nelson inputs 389 words in 19 minutes. What is his inputting rate?

   A about 20 words per minute

   B about 30 words per minute

   C about 40 words per minute

   D about 60 words per minute

5. There are 2.54 centimeters in 1 inch. What is a good estimate of the number of centimeters in 5 inches?

   A between 8 and 10 centimeters

   B between 10 and 12 centimeters

   C between 12 and 14 centimeters

   D between 14 and 16 centimeters

6. The scale on a blueprint of a house is $\frac{1}{2}$ inch:1 foot. The living room of the house is 20 feet long. How many inches long is the room in the drawing?

   Record your answer. Then fill in the bubbles. Be sure to use the correct place value.

⓪	⓪
①	①
②	②
③	③
④	④
⑤	⑤
⑥	⑥
⑦	⑦
⑧	⑧
⑨	⑨

NOTICE: Photocopying any part of this book is forbidden by law.

Lesson 15: **Ratios, Rates, and Proportions**

7. A map scale is 0.5 inch:20 miles. What is the actual distance between two cities that are 3 inches apart on the map? (Hint: Try the *draw a diagram* strategy.)

8. The last time Mr. Dali traveled to Canada, the value of 15 U.S. dollars was about 22 Canadian dollars. What was the value in U.S. dollars of 110 Canadian dollars? Explain. (Hint: Try the *write an equation* strategy.)

_____

_____

_____

9. Use the Venn diagram below. What is the ratio of the number of students in both clubs to all students in the two clubs? (Hint: Try the *use logical thinking* strategy.)

Academic Clubs

Theater Club 22 — 8 — Camera Club 18

10. The lengths of the sides of a square are doubled. Write a ratio that compares the area of the smaller square to the area of the larger square.

11. This table shows a set of scale distances and actual distances for a map.

Map Scale

Scale Distance	Actual Distance
0.2 inch	0.5 kilometer
0.4 inch	1 kilometer
0.6 inch	1.5 kilometers

What is the scale distance for an actual distance of 2.5 kilometers?

12. Glenn ran 34 kilometers in 4 hours, and Karl ran 26 kilometers in 3 hours. Who ran at a faster rate? Explain how you found the answer.

Strategies: Diagram • Model • Organized List • Table/Graph • Pattern • Predict & Test
Logical Thinking • Work Backward • Simpler Problem • Formula or Equation

Part II: Applying the Strategies—Solving Problems with Rational Numbers

16 Operations with Percents

Percents are often used to show and compare data. You can use percents to solve problems involving discounts, sales tax, interest earned, and finance charges.

Example 1 Jefferson Middle School soccer team plays in the Banana League. What percent of its games did the Jefferson soccer team win?

Soccer Team	Wins	Losses
Randolph Middle School	10	28
Jefferson Middle School	25	15
King Middle School	15	25

Find Out You want to find what percent of games was won. There is too much information. You need to know only that the Jefferson team won 25 games and lost 15 games.

Plan You can use the *write an equation* strategy to solve this problem.

Solve games won + games lost = total games played

25 + 15 = 40 *The team played 40 games.*

25 ÷ 40 = 0.625 *Divide games won by total games.*

0.625 × 100 = 62.5% *Multiply by 100 to find the percent.*

Solution: The Jefferson Middle School team won 62.5% of its games.

Look Back You can check the answer by changing the fraction $\frac{25}{40}$ to a percent.

$\frac{25}{40}$ = 0.625; 0.625 × 100 = 62.5% ✓

Try It

Example 2 Shanell gave her sister 50% of her CD collection. She gave 20% of the remaining CDs to her cousin. How many CDs did Shanell start with if she has 8 left?

Find Out You know how many CDs Shanell gave away. You want to find the number of CDs Shanell started with.

Plan Since you know how many CDs Shanell has left, you can use the *work backward* strategy to find how many CDs she started with.

Solve Shanell had 8 CDs left. This is 80% of the CDs she had after giving CDs to her sister.

Think: 100% − 20% = 80%

To find how many CDs Shanell had before giving some to her cousin, write an equation.

$8 = 80\% \times n$

$n = \underline{\hspace{1cm}}$

That number of CDs, *n*, is 50% of the number Shanell started with.

$\underline{\hspace{1cm}} = 50\% \times$ the number Shanell started with.

$\underline{\hspace{1cm}} = 50\% \times s$

$s = \underline{\hspace{1cm}}$

Solution: Shanell started with _____ CDs.

Look Back What other way can you check your solution?

_____

Use your method. Is your answer the same?

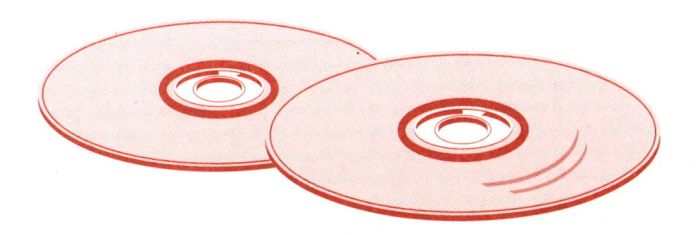

_____

Practice Test Questions
Find Out • Plan • Solve • Look Back

1. Last year, there were 396 seventh graders at Broadway Middle School. This year, there are 360 seventh graders. What was the percent of decrease? (Hint: Try the *write an equation* strategy.)

   A 9%
   B 10%
   C 91%
   D 110%

2. Desiree bought a pair of shoes with 40% of her money. She spent 75% of the remaining money on a computer game. Then she had $13.20 left. Determine which question could **not** be answered with the information provided. (Hint: Try the *work backward* strategy.)

   A How much did Desiree start with?
   B How much sales tax did she pay?
   C How much did the game cost?
   D How much did the shoes cost?

3. Eric bought a jacket on sale. If the original price of the jacket was $89.98, what was the sale price? (Hint: Try the *write an equation* strategy.)

   Jackets 20%
   Jeans 25%

   A $18.00
   B $22.50
   C $67.48
   D $71.98

4. In 2002, 2 U.S. governors were women. In 2003, the number was 6. By what percent did the number of women governors increase between 2002 and 2003?

   A 12%
   B 24%
   C 50%
   D 200%

5. José has saved 75% of the money he needs for a bicycle. If the bicycle costs $200, how much has he saved?

   A $25
   B $50
   C $150
   D $200

6. Tara's DVD player cost $130. Brad paid 30% less for his. How much, in dollars, did Brad's DVD player cost?

   Record your answer. Then fill in the bubbles. Be sure to use the correct place value.

*Lesson 16: **Operations with Percents***

7. A swimming pool is 35% full. The pool holds 15,000 gallons. How many gallons of water are in the pool? (Hint: Try the *look for a pattern* strategy.)

8. Karla made a large cube from 64 small cubes. Then she painted the outside faces of the large cube green. What percent of the small cubes has 3 green faces? (Hint: Try the *make a model* strategy.)

9. Students from Witherspoon School are planning a Games Day in an area that measures 20 feet by 50 feet. The beanbag toss takes 15 feet by 10 feet. Milk carton bowling takes 15 feet by 15 feet. The 3-legged race takes 10 feet by 25 feet, and the jump-rope contest takes 10 feet by 10 feet. The rest of the area will be used for playing board games. What percent of the area will be used for playing board games? (Hint: Try the *draw a diagram* strategy.)

10. Barry divided his collection of 200 baseball cards among his friends. Kendra got 15% of the cards, Manny got 10%, Serena got 8%, and Therese got the rest. How many cards did Therese get? What strategy did you use?

11. The perimeter is 48 centimeters. The width is 50% of its length. What is the length and width of the rectangle? Explain how you got your answer.

_____
_____
_____
_____

12. Marci, Brenda, Linda, Clyde, Kevin, and Tom are the members of the school chess team. For practice, each person will play every other person once. Explain in words and symbols how to determine what percent of the games is between a boy and a girl?

_____
_____
_____
_____

Strategies: Diagram • Model • Organized List • Table/Graph • Pattern • Predict & Test
Logical Thinking • Work Backward • Simpler Problem • Formula or Equation

Part II: Applying the Strategies—Solving Problems with Geometry and Measurement

17 Metric and Customary Units of Measurement

It is important to know how to compare units within the metric and customary measurement systems. It is also useful to be able to approximate conversions between the two systems.

Example 1

Tracy exercises by walking every day. She recorded distances in miles and kilometers from her pedometer. On Wednesday, she forgot to record the number of kilometers for 10 miles.

	Mon	Tue	Wed	Thu
Miles	4	8	10	12
Kilometers	6.4	12.8	?	19.2

About how many kilometers are equivalent to 10 miles?

Find Out

You know these equivalences.
4 miles = 6.4 kilometers
8 miles = 12.8 kilometers
12 miles = 19.2 kilometers

You need to find the number of kilometers equivalent to 10 miles.

Plan

You can use the *write an equation* and *look for a pattern* strategies to solve this problem. The sequence of numbers of miles Tracy walked (4, 8, 10, 12) does not have a regular pattern. Look for a relationship in the ratio of known miles to known kilometers.

Solve

$\frac{6.4}{4} = 1.6$ → $4 \times 1.6 = 6.4$

$\frac{12.8}{8} = 1.6$ → $8 \times 1.6 = 12.8$

$\frac{19.2}{12} = 1.6$ → $12 \times 1.6 = 19.2$

So, miles $\times$ 1.6 = kilometers.
$10 \times 1.6 = 16$

Solution: 10 miles is about 16 kilometers.

Look Back

To check your answer, you could write and solve a proportion such as:

$\frac{12.8}{8} = \frac{x}{10}$

$8x = 128$

$x = 16$ ✓

NOTICE: Photocopying any part of this book is forbidden by law.

Lesson 17: Metric and Customary Units of Measurement

Try It

Example 2

One day, Calvin recorded these four temperatures in both Fahrenheit (F) and Celsius (C) degrees.

Fahrenheit	32°	41°	50°	59°
Celsius	0°	5°	10°	15°

What is a good approximation of the number of degrees Fahrenheit that would be equivalent to 7° Celsius?

Find Out   You know the equivalencies in the table. You need to find about how many degrees Fahrenheit are equivalent to 7°C.

Plan   You can use the *make a graph* strategy to solve this problem. Make a coordinate graph. Label the horizontal axis *Degrees Celsius* and the vertical axis *Degrees Fahrenheit*.

Solve   What ordered pairs can you graph? _____

Graph the points and draw a line through them.

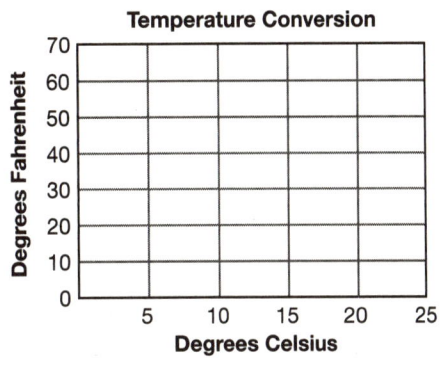

Temperature Conversion

[graph with x-axis "Degrees Celsius" from 0 to 25, y-axis "Degrees Fahrenheit" from 0 to 70]

Locate 7 on the horizontal axis. What does this represent?

_____

Draw a vertical line from 7 that intersects the line. What is the approximate *y*-coordinate of the point of intersection? _____

Solution: 7°C ≈ _____ °F

Look Back   You can verify your answer by using the formula that relates Fahrenheit degrees to Celsius degrees.

$$F = \frac{9}{5}(C + 32)$$

How does the result from the formula compare to the result from graphing?

_____

NOTICE: Photocopying any part of this book is forbidden by law.

Practice Test Questions
Find Out • Plan • Solve • Look Back

1. Leroy made a fruit punch using 120 milliliters of cranberry juice and 380 milliliters of apple juice. How many liters of punch did he make? (Hint: Try the *write an equation* strategy.)

1 milliliter = 0.001 liter

   A 0.5   C 50
   B 5   D 500

2. Guy needs to cut a piece of wood $2\frac{1}{2}$ feet long. He knows that 1 foot = 30.48 centimeters. How long should the piece of wood be in centimeters? (Hint: Try the *write a formula or an equation* strategy.)

   A 30.48
   B 60.96
   C 76.2
   D 762

3. Frannie bought 10 bags of popcorn, each with a mass of 30 grams. Which question **cannot** be answered with this information? (Hint: Try the *use logical thinking* strategy.)

   A How many grams of popcorn are there in all?
   B How many milligrams are in each bag?
   C How many grams of popcorn are in 3 bags?
   D How many days will it take Frannie to eat the popcorn?

4. A liter is about 1.1 quarts. About how many quarts of soup are equivalent to 15 liters of soup?

   A between 13 and 15
   B between 16 and 18
   C between 19 and 21
   D Not here

5. These are the masses of books in Naomi's backpack.

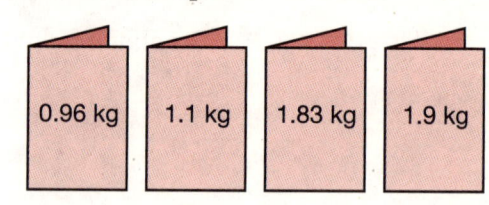

   0.96 kg   1.1 kg   1.83 kg   1.9 kg

   What is a good estimate of the total mass of the books?

   A 6 grams   C 4,000 grams
   B 3,000 grams   D 6,000 grams

6. A rectangular room measures 15 feet by 18 feet. How many square yards of carpeting are needed to cover the floor?

   Record your answer. Then fill in the bubbles. Be sure to use the correct place value.

Lesson 17: **Metric and Customary Units of Measurement**

7. This table shows some approximate equivalent measurements in feet and in meters.

Feet	10	20	23
Meters	3	6	7

About how many meters are equivalent to 13 feet? Explain how you found your answer. (Hint: Try the *make a table or graph* strategy.)

_____

_____

_____

8. What is the temperature in degrees Celsius when the temperature is 77°F? (Hint: Try the *write a formula or an equation* strategy.)

$$C = \tfrac{5}{9}(F - 32)$$

9. Delilah has five pieces of ribbon that are 4, 6, 8, 10, and 12 inches long. In how many ways can she choose 3 pieces of ribbon having a total length of more than 2 feet? (Hint: Try the *make an organized list* strategy.)

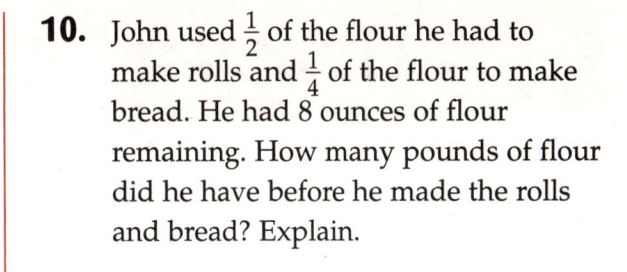

10. John used $\tfrac{1}{2}$ of the flour he had to make rolls and $\tfrac{1}{4}$ of the flour to make bread. He had 8 ounces of flour remaining. How many pounds of flour did he have before he made the rolls and bread? Explain.

_____

_____

_____

_____

11. The area of Pia's square poster is 0.64 square meter. What are the dimensions of her poster?

12. Gary has 5 meters of tape. He has a project that requires 15 feet of tape. Does he have enough tape? Explain.

1 meter is about 3.28 feet.

_____

_____

_____

Strategies: Diagram • Model • Organized List • Table/Graph • Pattern • Predict & Test
Logical Thinking • Work Backward • Simpler Problem • Formula or Equation

Part II: Applying the Strategies—Solving Problems with Geometry and Measurement

18 Lines and Angles

An understanding of lines and angles is key to solving problems that involve geometric figures.

Example 1

Find Out

Shari drew line *RS* and line *PQ* so that they intersect at point *T*. The measure of angle *RTP* is 55°. What is the measure of angle *PTS*?

You know that the two lines intersect. When two lines intersect, angles are formed. You know that one of the angles formed has a measure of 55°. You need to find the measure of one of the other angles.

Plan

You can use the *draw a diagram* strategy to solve this problem.

Solve

Draw lines *RS* and *PQ* intersecting at point *T*. Since ∠*RTP* has measure of 55°, make that angle acute in the drawing.

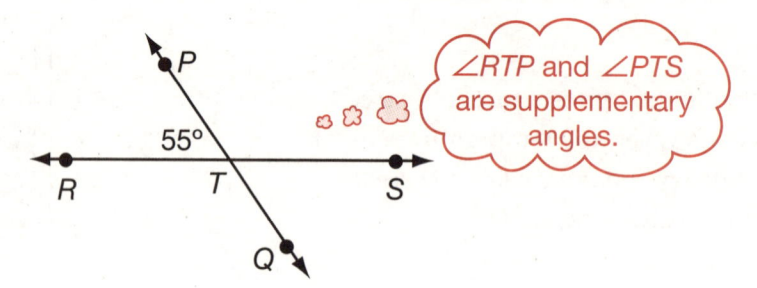

(∠RTP and ∠PTS are supplementary angles.)

measure of ∠*RTP* + measure of ∠*PTS* = 180°
measure of ∠*PTS* = 180° − measure of ∠*RTP*
= 180° − 55°
= 125°

Solution: The measure of ∠*PTS* = 125°.

Look Back

When two lines intersect, two pairs of congruent vertical angles are formed, so you can label angles *RTQ* and *STQ* with their measures.

To check your answer, add all four angle measures. The sum should be 360°.

measures of
∠*RTP* + ∠*PTS* + ∠*STQ* + ∠*RTQ* = 360°
55° + 125° + 55° + 125° = 360° ✔

56   NOTICE: Photocopying any part of this book is forbidden by law.

Lesson 18: **Lines and Angles**

Try It

Example 2 Angle A and angle B are complementary angles. The measure of ∠A is 5 times the measure of ∠B. What are the measures of ∠A and ∠B?

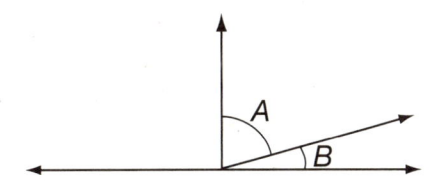

Find Out You know that the angles are complementary and that the measure of ∠A is 5 times the measure of ∠B. You need to find the measures of both angles.

Plan You can use the *write a formula or an equation* strategy to solve this problem.

Solve **Think:** What equation can you write to show that the angles are complementary?

measure of ∠A + measure of ∠B = _____

What expression represents the measure of ∠A? Let x represent the measure of ∠B.

measure of ∠A = ___x

measure of ∠A + measure of ∠B = 90°

x + ___x = 90° *Write an equation.*

___x = 90° *Solve for x.*

x = _____

measure of ∠A = 5 × measure of ∠B.

= 5 × _____

= _____

Solution: The measure of ∠A is _____, and the measure of ∠B is _____.

Look Back To check your answer, you can add the angle measures you found for ∠A and ∠B.

_____ + _____ = _____

Is the sum 90°? _____

Practice Test Questions
Find Out • Plan • Solve • Look Back

1. Line *RS* intersects line segment *AB* at its midpoint *M*. What conclusion can you draw from this information? (Hint: Try the *draw a diagram* strategy.)

   A MR = MS
   B AM = MS
   C AM = MB
   D AB = RS

2. Angle *D* and angle *F* are supplementary. The measure of angle *F* is 4 times the measure of angle *D*. What is the measure of angle *D*? (Hint: Try the *write an equation* strategy.)

   A 36°   C 90°
   B 45°   D 144°

3. Which equation can be used to find the measure of $\angle SOT$? (Hint: Try the *use logical thinking* strategy.)

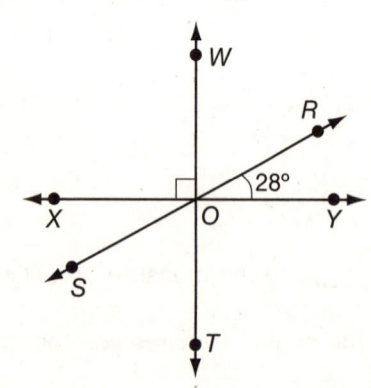

   A measure of $\angle SOT - 28° = 90°$
   B measure of $\angle SOT + 28° = 90°$
   C measure of $\angle SOT + 28° = 180°$
   D measure of $\angle SOT + 90° = 180°$

4. A geometric figure has an endpoint *J* and continues through point *K* without end. What is a name for this figure?

   A line *JK*
   B segment *JK*
   C ray *KJ*
   D ray *JK*

5. Angle *N* and angle *M* are supplementary angles, and the measure of angle *N* is greater than 65°. Which statement describes the possible measures of angle *M*?

   A measure of $\angle M < 25°$
   B measure of $\angle M < 65°$
   C measure of $\angle M > 115°$
   D measure of $\angle M < 115°$

6. $\angle T$ and $\angle V$ are supplementary angles. The measure of $\angle T$ is $3\frac{1}{2}$ times the measure of $\angle V$. What is the measure, in degrees, of $\angle V$?

   Record your answer. Then fill in the bubbles. Be sure to use the correct place value.

Lesson 18: **Lines and Angles**

7. Sunny drew angles with these measures.

17°, 22°, 44°, 46°, 54°, 58°, 68°, 73°, 83°

How many pairs of complementary angles are there? (Hint: Try the *make an organized list* strategy.)

8. Michael draws line *AB* parallel to line *CD*. What fraction of the angles in this figure has a measure of 35°? Explain. (Hint: Try the *use logical thinking* strategy.)

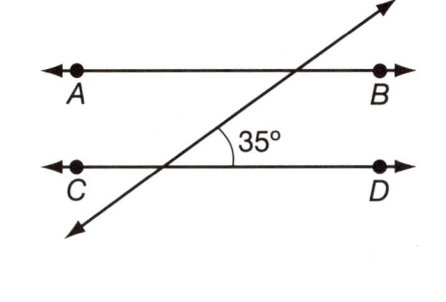

_____
_____
_____

9. Jody says it is possible for an angle to be complementary and supplementary to the same angle. Erin says it is not possible. Who is correct? Explain. (Hint: Try the *use logical thinking* strategy.)

_____
_____
_____
_____

10. Angles *ABC* and *DEF* are complementary angles. Angles *DEF* and *GJK* are supplementary angles. The measure of angle *GJK* is 125°. What is the measure of angle *ABC*?

11. Is it possible for two obtuse angles to be a supplementary pair of angles? Justify your answer.

_____
_____
_____

Strategies: Diagram • Model • Organized List • Table/Graph • Pattern • Predict & Test
Logical Thinking • Work Backward • Simpler Problem • Formula or Equation

Part II: Applying the Strategies—Solving Problems With Geometry and Measurement

19 Polygons and Circles

Understanding the properties of polygons, knowing how to classify polygons, and finding the area, perimeter, and circumference of plane figures are important skills for solving problems.

Example 1

Rosalie drew this Venn diagram to classify quadrilaterals. The possible labels for regions A, B, C, and D are *Parallelograms*, *Rectangles*, *Rhombuses*, and *Squares*. Which label will she put in each region?

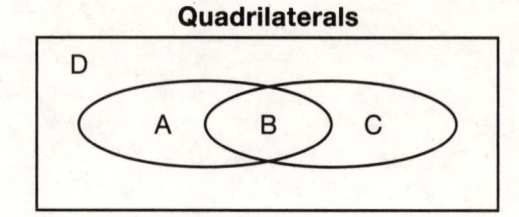

Quadrilaterals

Find Out You know the possible labels for regions A, B, C, and D. You need to find the correct region for each label.

Plan You can use the use *logical thinking* strategy to solve this problem.
Think: How are quadrilaterals, parallelograms, rectangles, rhombuses, and squares related?

Rectangle

Rhombus

Square

Solve Rectangles and rhombuses are parallelograms. Squares are rectangles. So, squares are parallelograms.
Label region D *Parallelograms*.
Since squares are rectangles, label region A *Rectangles* and region B *Squares*.
Since squares are also rhombuses, label region C *Rhombuses*.

Solution:

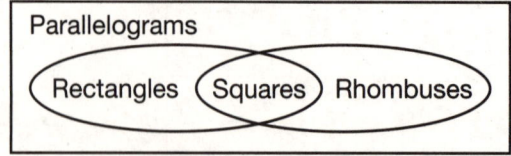

Quadrilaterals
Parallelograms
Rectangles — Squares — Rhombuses

Look Back Check your labels by comparing the definitions of each type of quadrilateral to your Venn diagram.
- A *parallelogram* is a quadrilateral with opposite sides parallel. ✔
- A *rectangle* is a parallelogram with right angles. ✔
- A *rhombus* is a parallelogram with four congruent sides. ✔
- A *square* is a rectangle with four congruent sides. ✔

NOTICE: Photocopying any part of this book is forbidden by law.

Lesson 19: **Polygons and Circles**

Try It

Example 2

The formula for finding the area of a circle is:

Area = π × radius²

When the radius of a circle is doubled, what happens to the area?

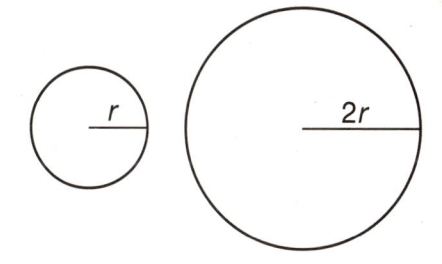

Find Out

You know that the formula for the area of a circle involves the radius. You want to find out what effect doubling the radius has on the area.

Plan

You can use the *predict and test* strategy to solve this problem. Pick a radius and double it. Find the areas of the circles and compare.

Solve

Try a radius of 2 and a radius 4. What is the area of each circle?

Area = π × radius² 　　　　Area = π × radius²
　　 = 3.14 × 2²　　　　　　　　 = 3.14 × 4²
　　 = 3.14 × 4 = 12.56　　　　 = 3.14 × 16 = 50.24

How do they compare?

50.24 ÷ 12.56 = 4

> The radius was multiplied by 2. The area is multiplied by 4.

Double the radius again. Try a radius of 8. What is the area?

_____

How does the area compare to the circle with a radius of 4?

_____

Double the radius again. Try a radius of 16. What is the area?

_____

How does the area compare to the circle with a radius of 8?

_____

What pattern do you see? _____

Solution: When the radius of a circle is doubled, the area _____

_____

Look Back

How can you check your answer? Show your work.

_____

NOTICE: Photocopying any part of this book is forbidden by law.

Practice Test Questions
Find Out • Plan • Solve • Look Back

1. Which of the following is **not** a true statement about the relationship of triangles? (Hint: Try the *use logical thinking* strategy.)

   A An isosceles triangle can be a right triangle.

   B An equilateral triangle can be a right triangle.

   C An equilateral triangle is also an isosceles triangle.

   D A triangle can be acute and isosceles.

2. When you double the length and width of a rectangle, what happens to the perimeter? (Hint: Try the *predict and test* strategy.)

   A It becomes twice as great.

   B It becomes four times as great.

   C It becomes half as great.

   D It becomes one fourth as great.

3. Which set of angle measures could be the angle measures of a quadrilateral? (Hint: Try the *work backward* strategy.)

   A 30°, 60°, 90°, 90°

   B 40°, 40°, 50°, 50°

   C 75°, 110°, 115°, 90°

   D 85°, 95°, 93°, 87°

4. A circular table has a circumference of 314 centimeters. What is the diameter of the table?

   A 5 centimeters

   B 10 centimeters

   C 100 centimeters

   D 1,000 centimeters

5. Monty is mowing lawns. One lawn is circular with a diameter of 10 yards. What is the area he has to mow?

   A 31.4 square yards

   B 62.8 square yards

   C 78.5 square yards

   D 314 square yards

6. A carpet designer wants to cover an 8-foot by 12-foot rectangular rug with circles. How many circles with a radius of 1 foot will fit on the carpet if the circles don't overlap?

   Record your answer. Then fill in the bubbles. Be sure to use the correct place value.

Lesson 19: Polygons and Circles

7. In a certain square, the number of inches in the perimeter is the same as the number of square inches in the area. What is the measure of a side of this square? (Hint: Try the *predict and test* strategy.)

8. What is the area of the shaded region? (Hint: Try the *write a formula or an equation* strategy.)

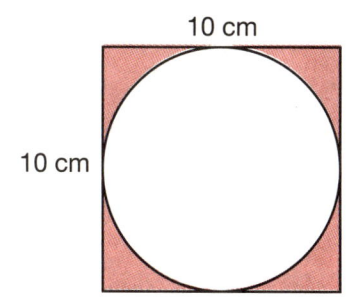

10 cm
10 cm

9. Ted is using square tiles measuring one inch on a side to make a figure. Each tile shares at least one full side with another tile. If he uses 9 tiles, what is the greatest possible perimeter his figure could have? (Hint: Try the *make a model* strategy.)

10. Carlton's table shows the sum of the interior angle measures for polygons with different numbers of sides.

Number of sides	3	4	5	6
Interior Angle Sum	180°	360°	540°	720°

What is the measure of each interior angle in a regular polygon with 9 sides?

11. How many triangles are formed when you draw all the diagonals from one vertex of a regular octagon?

12. This track has the shape of a rectangle with a semicircle at each end. What is the perimeter of the track?

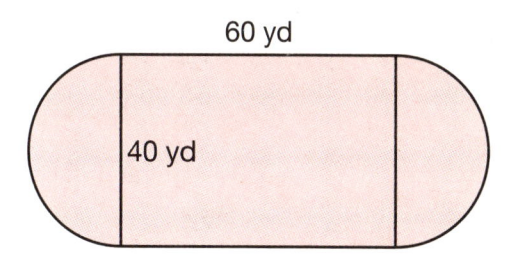

60 yd
40 yd

Strategies: Diagram • Model • Organized List • Table/Graph • Pattern • Predict & Test
Logical Thinking • Work Backward • Simpler Problem • Formula or Equation

Part II: Applying the Strategies—Solving Problems With Geometry and Measurement

20 Congruent and Similar Figures

Congruent figures are the same shape and size. *Similar figures* are the same shape, but not necessarily the same size. A *scale* is a ratio that compares the sizes of two similar shapes.

Example 1

Brock made this drawing of the two triangular gardens in a neighborhood park. Triangles *ABC* and *DEF* are similar.

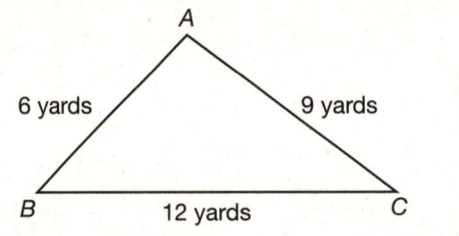

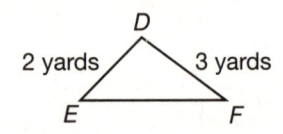

What is the length of side *EF* in triangle *DEF*?

Find Out

You know that the figures are similar. You also know that in similar figures, the ratios of the lengths of the corresponding sides are equal. You need to find the missing length in triangle *DEF*.

Plan

You can use the *write a formula or an equation* strategy. In this case, the equation you use is a proportion.

Solve

Find the ratio of the pairs of corresponding sides.

$\frac{AB}{DE} = \frac{6}{2} = \frac{3}{1}$

$\frac{AC}{DF} = \frac{9}{3} = \frac{3}{1}$

$\frac{BC}{EF} = \frac{12}{EF}$

Write and solve a proportion.

$\frac{12}{EF} = \frac{3}{1}$

$3 \times EF = 12 \times 1$ *Write the cross products.*

$EF = 4$ *Solve for EF.*

Solution: The length of side *EF* = 4 yards.

Look Back

You can work backward to check your answer. Substitute 4 in the proportion $\frac{12}{EF} = \frac{3}{1}$.

$\frac{12}{4} = \frac{3}{1}$

$12 = 12$ ✔

Try It

Example 2

The coordinates of rectangle JKLM are J(1,3), K(4,3), L(4,1), and M(1,1). In a congruent rectangle, J' corresponds to J and K' corresponds to K. J' is (3, 6) and K' is (6, 6). What are the coordinates of L' and M'?

Find Out

You know the coordinates of the vertices of rectangle JKLM and of a rectangle that is congruent to JKLM.

You need to find the coordinates of L' and M' in the congruent rectangle.

Plan

You can use the *make a graph* strategy to solve this problem.

Solve

Graph rectangle JKLM. Then graph the two vertices of the congruent rectangle and connect them with a line segment.

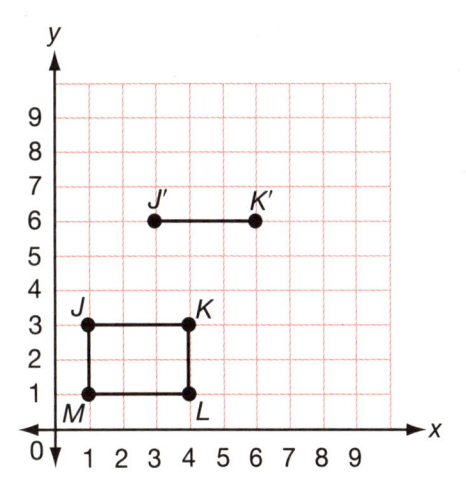

How must the shape and size of the congruent rectangle compare with the shape and size of rectangle JKLM?

_____

How can you find the location of L' and M'?

_____

Draw segments to complete the rectangle.

Solution: The coordinates of L' are _____ and M' are _____.

Look Back

Compare the rectangles. Are the corresponding sides congruent? Are the corresponding angles congruent?

_____

Practice Test Questions
Find Out • Plan • Solve • Look Back

1. Alexa drew these two similar rectangles. What is the length of the side labeled *x*? (Hint: Try the *write an equation* strategy.)

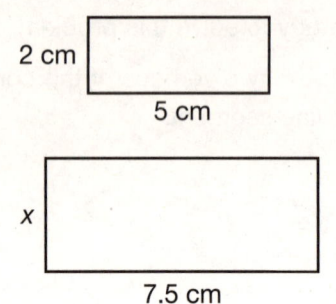

2 cm
5 cm

x
7.5 cm

  A 3 centimeters
  B 3.5 centimeters
  C 4 centimeters
  D 4.5 centimeters

2. Triangle *GHJ* is congruent to triangle *PQR*. Triangle *GHJ* has vertices *G*(1,1), *H*(4,2), and *J*(4,0). Triangle *PQR* has vertices *Q*(5,5) and *R*(5,3). Which could be coordinates of vertex *P*? (Hint: Try the *make a graph* strategy.)

  A (2,3)   **C** (2,5)
  B (2,4)   **D** (5,2)

3. Triangle *ABC* is congruent to triangle *HKL*. Angle *A* corresponds to angle *H*, and angle *B* corresponds to angle *K*. What side in triangle *HKL* corresponds to side *AC* in triangle *ABC*? (Hint: Try the *draw a diagram* strategy.)

  A HK   **C** KH
  B KL   **D** HL

4. Lydia graphs rectangle *RSTV* on a coordinate grid. The rectangle is 2 units wide and 3 units high. Which could be the coordinates of a rectangle congruent to *RSTV*?

  A (1,1), (3,1), (3,3), (1,3)
  B (1,1), (4,1), (4,2), (1,2)
  C (2,1), (4,1), (4,4), (2,4)
  D (1,1), (3,1), (3,5), (1,5)

5. A scale drawing of a house has a scale $\frac{1}{4}$ inch:1 foot. The actual kitchen is 16 feet long. How long is the kitchen in the drawing?

  A 2 inches   **C** 8 inches
  B 4 inches   **D** 16 inches

6. Right triangle *XYZ* is congruent to right triangle *RST*. Right angle *X* corresponds to right angle *R*, and angle *Y* corresponds to angle *S*. If angle *Y* measures 38°, what is the measure, in degrees, of angle *T*?

Record your answer. Then fill in the bubbles. Be sure to use the correct place value.

Lesson 20: **Congruent and Similar Figures**

7. A hallway on a blueprint of a house is 5 inches long. The scale of the blueprint is $\frac{1}{4}$ inch = 1 foot. What is the actual length of the hallway? Explain. (Hint: Try the *write an equation* strategy.)

_____

_____

_____

8. Claudia says that if two triangles are congruent, they are also similar. Linton says that if two squares are similar, they are also congruent. Are they both correct? Explain. (Hint: Try the *draw a diagram* strategy.)

_____

_____

_____

9. Raul used a photocopier to enlarge a picture of a triangle to 1.5 times its original size. If the base of the triangle in the enlargement was 18 centimeters, what was the base in the original triangle? (Hint: Try the *work backward* strategy.)

10. Elijah drew and labeled these two similar triangles.

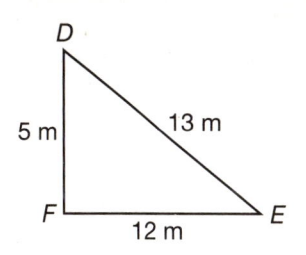

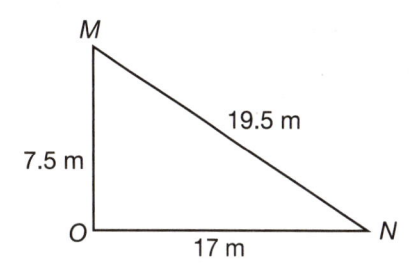

Triangle DFE: DF = 5 m, DE = 13 m, FE = 12 m

Triangle MON: MO = 7.5 m, MN = 19.5 m, ON = 17 m

Which segment on triangle MON did he label with an incorrect length? What should the label be?

11. As part of an art project, Julie is making a set of similar triangles. The table shows the side lengths of her first three triangles. If the pattern continues, what will be the side lengths of her fifth triangle?

Triangle	AB	BC	CA
1	6	8	10
2	9	12	15
3	12	16	20

Strategies: Diagram • Model • Organized List • Table/Graph • Pattern • Predict & Test
Logical Thinking • Work Backward • Simpler Problem • Formula or Equation

Part II: Applying the Strategies—Solving Problems With Geometry and Measurement

21 Rotations, Reflections, and Translations

Rotations, reflections, and translations are types of transformations. A *rotation* is a turn, a *reflection* is a flip, and a *translation* is a slide.

Example 1

Sylvie is making a design on a coordinate grid. She made the first part of her design in the first quadrant. She would like to continue the design by drawing a reflection of the first part across the *y*-axis. What will be the coordinates of the vertices of the reflection?

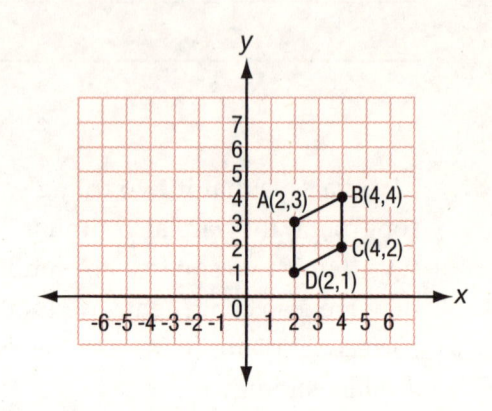

Find Out   You know the coordinates of the vertices of the figure in the first quadrant. You need to find the coordinates of the reflection of each vertex across the *y*-axis.

Plan   You can use the *draw a diagram* strategy to solve this problem.

Solve
- Find the distance of each vertex from the *y*-axis.
- Then find the reflection of each vertex by locating the point with the same *y*-coordinate that is the same number of units from the *y*-axis but in the second quadrant.
- Connect the vertices and compare the figures.

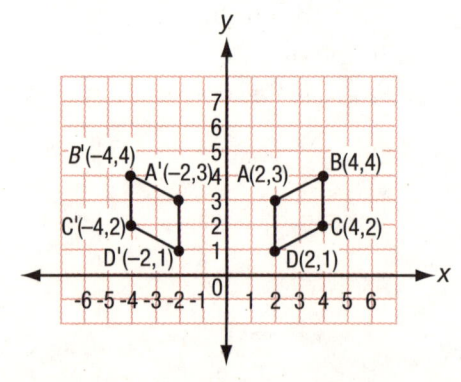

Solution: The coordinates of the vertices of the reflection are $A'(-2,3)$, $B'(-4,4)$, $C'(-4,2)$, and $D'(-2,1)$.

Look Back   One way to check the reflection is to fold the graph along the *y*-axis and hold the folded paper up to the light. The figures should coincide. ✔

Lesson 21: **Rotations, Reflections, and Translations**

Try It

Example 2 Triangle P'Q'R' is the image of triangle PQR after PQR is translated 2 units to the right and 3 units down. What are the coordinates of the vertices of triangle PQR?

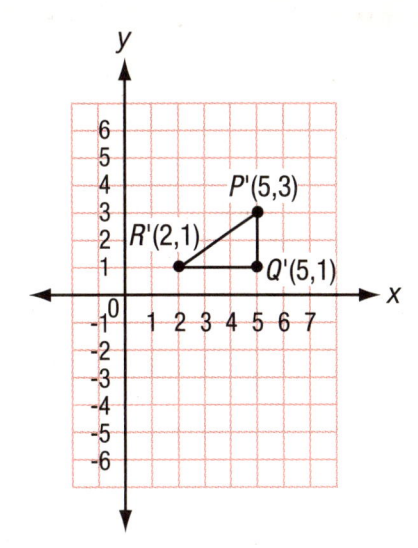

Find Out You know the coordinates of the image of triangle PQR. You know how triangle PQR was translated. You need to find the coordinates of triangle PQR.

Plan You can use the *work backward* strategy to solve this problem.

Solve How can you work backward from a translation of "2 units to the right"?

_____

How can you work backward from a translation of "3 units down"?

_____

Working backward from P'(5,3), the coordinates of P are

(5 − 2, 3 + 3), or (____, ____).

Working backward from Q'(5,1), the coordinates of Q are ____, ____.

Working backward from R'(2,1), the coordinates of R are ____, ____.

Solution: The coordinates of the vertices of triangle PQR are

P(____, ____), Q(____, ____), and R(____, ____).

Look Back How can you check your solution?

_____

NOTICE: Photocopying any part of this book is forbidden by law.

Mathematics Problem Solving Coach, Level G: Strategies and Applications

Practice Test Questions
Find Out • Plan • Solve • Look Back

1. Angelo draws point $G(3,4)$ in quadrant I. Then he reflects it across the y-axis. In which quadrant is the image of point G? (Hint: Try the *draw a diagram* strategy.)

    A Quadrant I
    B Quadrant II
    C Quadrant III
    D Quadrant IV

2. Triangle $A'B'C'$ is the image of triangle ABC after it was reflected across the x-axis. It has coordinates $A'(3,3)$, $B'(4,1)$, and $C'(1,2)$. What are the coordinates of triangle ABC? (Hint: Try the *work backward* strategy.)

    A $A'(3,-3), B'(4,-1), C'(1,-2)$
    B $A'(-3,3), B'(-4,1), C'(-1,2)$
    C $A'(-3,3), B'(-1,4), C'(-2,1)$
    D $A'(-3,-3), B'(-4,-1), C'(-1,-2)$

3. Point R is at $(3,7)$. It is rotated $90°$ clockwise around the origin. What are the coordinates of the image, R'? (Hint: Try the *write an equation* strategy.)

    A $R'(3,-7)$
    B $R'(7,3)$
    C $R'(7,-3)$
    D Not here

4. The shaded figure is the image of the unshaded figure after a transformation. What is the transformation?

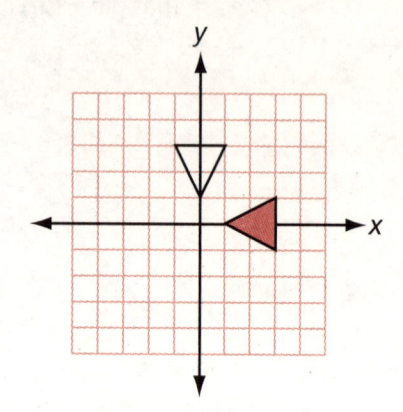

    A a reflection across the y-axis
    B a reflection across the x-axis
    C a quarter-turn rotation around the origin
    D a half-turn rotation around the origin

5. Melba translates point C 3 units to the right and 2 units down. The image of point C is $C'(24,19)$. What is the y-coordinate of point C?

Record your answer. Then fill in the bubbles. Be sure to use the correct place value.

Lesson 21: **Rotations, Reflections, and Translations**

6. Keesha puts her answer sheet in the upper left corner of her desk. Then she slides it over to the upper right corner. What kind of transformation is this? (Hint: Try the *make a model* strategy.)

7. What is the *x*-coordinate of the image of point *T*(6,5) after a reflection across the *x*-axis? Explain your answer. (Hint: Try the *draw a diagram* strategy.)

_____
_____
_____
_____

8. Bailey picks up a book and realizes that it is face up but upside down. What type of transformation does he use so that he can read his book? (Hint: Try the *make a model* strategy.)

9. This table shows the coordinates of points and their images after a half-turn about the origin.

Point	Image
R(3,4)	R'(−3,−4)
S(4,−3)	S'(−4,3)
T(2,5)	T'(−2,−5)
V(7,2)	V'(−7,−2)

What are the coordinates of the image of *P*(1,8) after a half turn about the origin? Explain.

_____
_____
_____

10. You write your name on a piece of paper and hold it up to a mirror. What type of transformation is the image in the mirror?

11. Vangie draws point *H* at (5,3). If she translates it 2 units right and 3 units up, what will be the coordinates of the image?

Strategies: Diagram • Model • Organized List • Table/Graph • Pattern • Predict & Test
Logical Thinking • Work Backward • Simpler Problem • Formula or Equation

Part II: Applying the Strategies—Solving Problems With Geometry and Measurement

22 Solid Figures

Knowing how to classify and measure solids can help you describe and measure the everyday objects.

Example 1 The net for a solid figure consists of 3 congruent rectangles and 2 congruent equilateral triangles. The side of a triangle equals the width of a rectangle. What solid figure can be built from this net?

Find Out You know the properties of the solid figure. You need to classify it.

Plan You can use the *make a model* and *predict and test* strategies to solve this problem.

Solve Since the 3 rectangles are congruent, try placing them side by side along their lengths.

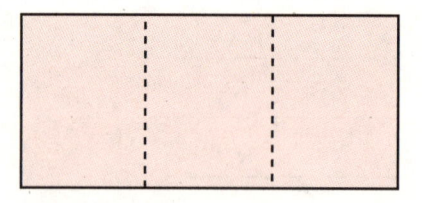

Since the 2 triangles have sides equal to the width of a rectangle, try placing triangles along two of the widths.

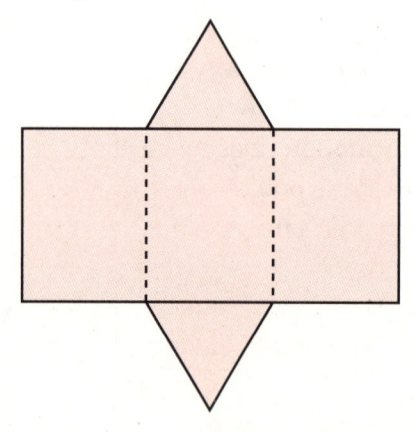

If you cut out and folded this net, you would have this solid.

Solution: The solid is a triangular prism.

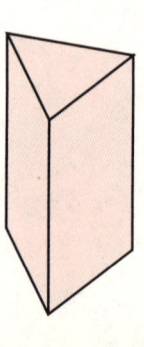

Look Back Try making another net with the given information. Your resulting solid figure should be a triangular prism. ✔

72   NOTICE: Photocopying any part of this book is forbidden by law.

Lesson 22: **Solid Figures**

Try It

Example 2

The pyramid at the Louvre Museum in Paris, France, was designed by the Chinese-born American architect, I. M. Pei. The approximate dimensions of this square pyramid are shown below. About what is the volume of this pyramid?

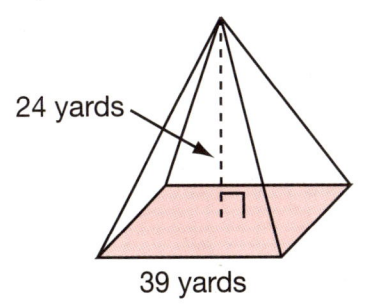

24 yards

39 yards

Find Out

You know that the figure is a pyramid with a square base. You know the length of a side of the square base and the height of the pyramid. You need to find the volume of the pyramid.

Plan

You can use the *write a formula* strategy to solve this problem.

Solve

The formula for the volume of a pyramid is

$$\text{Volume} = \frac{1}{3} \times \text{area of the base} \times \text{height}.$$

Substitute the values for the area of the base and the height in the formula.

$V = \frac{1}{3} \times B \times h$ ◁ The area of the base is 39 × 39 = 1,521 yards.

$V = \frac{1}{3} \times \underline{\qquad} \times \underline{\qquad}$

$V = \underline{\qquad}$

What unit will you use for the volume? _____

Solution: The volume of the pyramid at the Louvre Museum is about

_____ .

Look Back

Instead of multiplying by $\frac{1}{3}$ when computing the volume, you could divide by 3. Try solving the problem this way. Do you get the same answer?

_____

NOTICE: Photocopying any part of this book is forbidden by law.

Mathematics Problem Solving Coach, Level G: Strategies and Applications

Practice Test Questions
Find Out • Plan • Solve • Look Back

1. What will be the volume of the solid represented by this net? Each square measures 2 centimeters on a side. (Hint: Try the *make a model* strategy.)

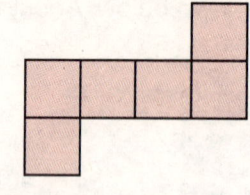

   A 4 cubic centimeters
   B 8 cubic centimeters
   C 16 cubic centimeters
   D 24 cubic centimeters

2. A water tank in the shape of a cylinder has a radius of 5 feet and a height of 10 feet. About what is the volume of the water tank? (Hint: Try the *write a formula* strategy.)

   A 157 cubic feet
   B 314 cubic feet
   C 785 cubic feet
   D 3,140 cubic feet

3. Wiley has a box that measures 12 inches by 12 inches by 9 inches. Which of the following questions **cannot** be answered with this information? (Hint: Try the *use logical thinking* strategy.)

   A How many pounds will the box hold?
   B What is the surface area of the box?
   C How many 3-inch cubes will the box hold?
   D What is the volume of the box?

4. What is the least amount of wrapping paper needed to cover a box that measures 18 inches by 15 inches by 3 inches?

   A 369 square inches
   B 630 square inches
   C 738 square inches
   D 810 square inches

5. The length of the edges of a cube is doubled. How does the volume of the larger cube compare to the volume of the smaller cube?

   A It is 2 times as great.
   B It is 3 times as great.
   C It is 4 times as great.
   D Not here

6. How many cubes, each 6 inches by 6 inches by 6 inches, will fit in a crate that measures 24 inches by 36 inches by 18 inches?

   Record your answer. Then fill in the bubbles. Be sure to use the correct place value.

Lesson 22: Solid Figures

7. What is the volume of this solid? Explain your answer. (Hint: Try the *write a formula* strategy.)

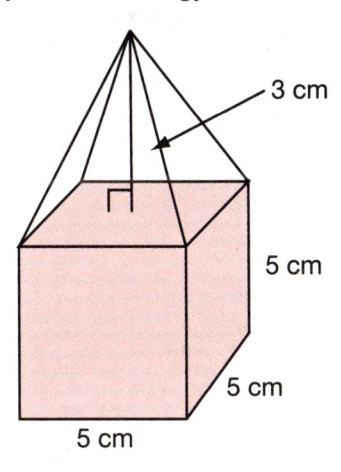

3 cm
5 cm
5 cm
5 cm

_____
_____
_____
_____

8. A prism has pentagons for bases. How many edges does the prism have in all? (Hint: Try the *make a model* strategy.)

9. A cone has a volume of 24 cubic feet. The area of its base is 18 square feet. What is the height of the cone? (Hint: Try the *write an equation* strategy.)

10. In how many ways can you select three dimensions of a box from the following so that the volume of the box is 48 cubic centimeters?

1 cm, 2 cm, 3 cm, 4 cm, 6 cm, 8 cm, 12 cm

11. This table shows edge lengths for cubes and corresponding volumes.

Edge	Volume
1 inch	1 cubic inch
2 inches	8 cubic inches
3 inches	27 cubic inches
4 inches	64 cubic inches

What is the edge length of a cube that has a volume of 343 cubic inches?

12. A rectangular prism has a volume of 216 cubic centimeters, and one of its dimensions is 3 centimeters. The other two dimensions are an odd number and an even number. One of the numbers is 8 times the other. What are the other dimensions? Explain.

_____
_____
_____
_____

Strategies: Diagram • Model • Organized List • Table/Graph • Pattern • Predict & Test
Logical Thinking • Work Backward • Simpler Problem • Formula or Equation

NOTICE: Photocopying any part of this book is forbidden by law.

Part II: Applying the Strategies—Solving Problems with Algebra

23 Compare and Order Rational and Irrational Numbers

A *rational number* can be expressed as the ratio of an integer to a nonzero integer. An *irrational number* cannot be expressed as the ratio of an integer to a nonzero integer.

Example 1

Anna and Betty are painting square designs with diagonals in a mural. Anna says that the diagonal in her square will be $\sqrt{128}$ feet long. Betty says that the diagonal in her square will be 11.5 feet long. Who will paint the longer diagonal?

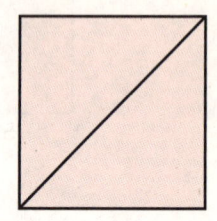

Find Out

You know two lengths, one expressed as an irrational number and one expressed as a rational number. You need to find which number is greater so that you can tell which diagonal is longer.

Plan

You can *use logical thinking* to solve this problem. Find a rational number approximation of $\sqrt{128}$. Then compare the approximation to 11.5.

Solve

Look for two radical expressions you can write as rational numbers, one that is less than $\sqrt{128}$ and one that is greater than $\sqrt{128}$.

$\left.\begin{array}{l}\sqrt{121} = 11 \\ \sqrt{128} = ? \\ \sqrt{144} = 12\end{array}\right\}$ *$\sqrt{128}$ is greater than 11 but less than 12.*

Since 128 − 121 = 7 and 144 − 128 = 16, 128 is closer to 121 than to 144, and $\sqrt{128}$ is closer to 11 than to 12.

So, $\sqrt{128} < 11.5$.

Solution: Betty will paint the longer diagonal.

Look Back

You can use a number line to check your answer.

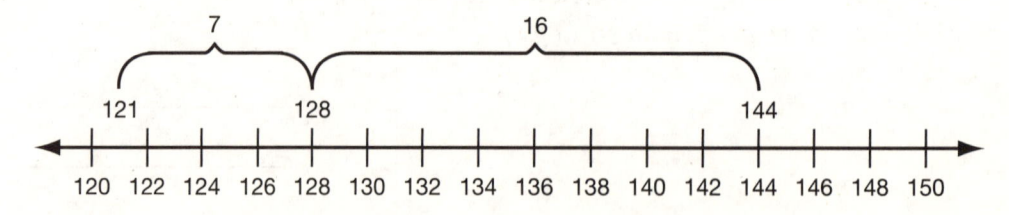

Since 128 is closer to 121 than to 144, $\sqrt{128}$ is closer to 11 than to 12.
So, $\sqrt{128} < 11.5$. ✔

Lesson 23: Compare and Order Rational and Irrational Numbers

Try It

Example 2 Shawna wrote this number pattern. What is the seventh term in the pattern?

$$\sqrt{144}, \sqrt{169}, \sqrt{196}, \sqrt{225}, \ldots$$

Find Out You have a sequence of four numbers in radical form. You need to find the seventh number.

Plan You can use the *look for a pattern* strategy to solve this problem.

Solve Look for a pattern in the differences of the numbers under the radical signs.

$$\sqrt{144} \quad \sqrt{169} \quad \sqrt{196} \quad \sqrt{225}$$

_____   _____   _____   ← *Difference of numbers under the radical sign.*

What will you add to $\sqrt{225}$ to find the fifth number in the pattern? _____

What will you add to the fifth number to find the sixth number? _____

What will you add to the sixth number to find the seventh number? _____

Solution: The seventh term in the pattern is _____.

Look Back You can write each radical as an integer.
Write the first four terms as integers. Continue the pattern to the seventh term.

_____, _____, _____, _____, _____, _____, _____

Write the seventh integer in radical form. _____

Is it the same as your answer above? _____

Mathematics Problem Solving Coach, Level G: Strategies and Applications

Practice Test Questions
Find Out • Plan • Solve • Look Back

1. Where should Marlon graph $-2\frac{1}{2}$ on a number line? (Hint: Try the *use logical thinking* strategy.)

   A between -2 and -1
   B between -2 and -3
   C between 2 and 3
   D between -2 and $-\frac{1}{2}$

2. Ruth wrote this number pattern. What is the next term in her pattern? (Hint: Try the *look for a pattern* strategy.)

   $2\frac{1}{2}, 1, -\frac{1}{2}, -2, \underline{\ ?\ }$

   A $-2\frac{1}{2}$
   B -3
   C $-3\frac{1}{2}$
   D Not here

3. Orlando lists the rational numbers in order from least to greatest.

   $3, -0.75, -1\frac{1}{4}, \frac{4}{5}$

   Which is correct? (Hint: Try the *draw a diagram* strategy.)

   A $\frac{4}{5}, -1\frac{1}{4}, -0.75, 3$
   B $-0.75, -1\frac{1}{4}, \frac{4}{5}, 3$
   C $-1\frac{1}{4}, -0.75, \frac{4}{5}, 3$
   D $3, \frac{4}{5}, -0.75, -1\frac{1}{4}$

4. What inequality is shown in Taylor's graph?

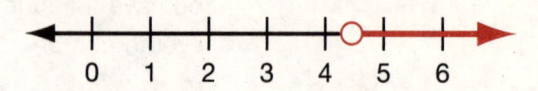

   A $x > \sqrt{4.5}$
   B $x > \sqrt{6}$
   C $x > \sqrt{10}$
   D $x > \sqrt{20}$

5. The school baseball coach recorded the batting averages of her four best hitters. Which list shows the batting averages from greatest to least?

   A 0.301, 0.290, 0.265, 0.256
   B 0.301, 0.290, 0.256, 0.265
   C 0.256, 0.265, 0.290, 0.301
   D 0.301, 0.256, 0.265, 0.290

6. Carol wants to glue a ribbon along the diagonal of a rectangular placemat. She calculates its length as $\sqrt{105}$ inches. To the nearest whole number, how many inches of ribbon should she buy?

   Record your answer. Then fill in the bubbles. Be sure to use the correct place value.

Lesson 23: **Compare and Order Rational and Irrational Numbers**

7. Amy draws two squares. One has sides that are 6 centimeters long and the other has sides that are $\sqrt{6}$ centimeters long. Which square is larger? Explain. (Hint: Try the *use logical thinking* strategy.)

_____

_____

_____

_____

8. Use >, =, or < to write a number sentence that compares 8 and $\sqrt{56}$. (Hint: Try the *solve a simpler problem* strategy.)

9. Four students recorded the amount of time they spent watching television over a weekend.

2.6 hours $2\frac{3}{4}$ hours

2.25 hours $2\frac{5}{12}$ hours

Write the numbers of hours from least to greatest. (Hint: Try the *draw a diagram* strategy.)

10. Vaughn wrote this number pattern. What is the sixth term in his pattern?

2.5, 1.5, 0.5, −0.5, ? , ?

11. A square lawn measures 10 feet long and 10 feet wide. How long is a diagonal path across the lawn? Write the answer in radical form and as an integer to the nearest whole number.

12. Heather wants to graph the solution of the inequality $x \leq \sqrt{12}$ on a number line. Show what her graph should look like.

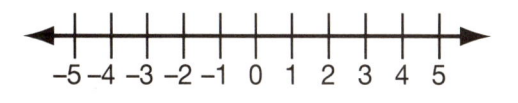

```
◄—+—+—+—+—+—+—+—+—+—+—►
 −5 −4 −3 −2 −1  0  1  2  3  4  5
```

Strategies: Diagram • Model • Organized List • Table/Graph • Pattern • Predict & Test
Logical Thinking • Work Backward • Simpler Problem • Formula or Equation

Part II: Applying the Strategies—Solving Problems With Algebra

24 Expressions, One-Step, and Two-Step Equations

To solve problems using expressions and equations, apply opposite operations and combine like terms.

Example 1

Karyn works after school delivering groceries. She earns a base salary of $8 per day plus $2 for each delivery she makes. On Monday, she earned $26. How many deliveries did she make?

Find Out

You know that Karyn earned $26. You know that her base salary is $8 per day and that she earns $2 for each delivery. You need to find the number of deliveries she made.

Plan

You can use the *make a table* strategy to solve this problem.

Solve

Make a table to show how Karyn's earnings increase with each delivery

Karyn's Earnings										
Number of Deliveries	0	1	2	3	4	5	6	7	8	9
Earnings	$8	$10	$12	$14	$16	$18	$20	$22	$24	$26

Solution: Karyn made 9 deliveries on Monday.

Look Back

You can check your answer by writing an equation.

Let d represent the number of deliveries.
Since Karyn makes $2 per delivery, $2d$ represents the amount she makes for all the deliveries.

Write the equation.
amount for all deliveries + base pay = total pay
$2d + 8 = 26$
$d = 9$ ✔

80   NOTICE: Photocopying any part of this book is forbidden by law.

Lesson 24: **Expressions, One-Step, and Two-Step Equations**

Try It

Example 2

An expression for how far an object falls when it is dropped from a high place is $16t^2$, where t represents the number of seconds the object has been falling. A pebble is dropped from a cliff. How many seconds does it take for the pebble to fall 576 feet?

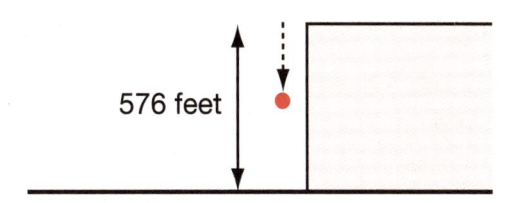

Find Out

You know that the pebble falls 576 feet. You know that the expression $16t^2$ gives the number of feet an object falls in t seconds. You need to find t, the number of seconds it takes for the pebble to fall 576 feet.

Plan

You can use the *predict and test* strategy to solve this problem.

Solve

Try different values for t in $16t^2$. Look for a value of t that will make $16t^2 = 576$.

t	$16t^2$	Too low or too high?
1	$16(1)^2 = 16 \times 1 = $ _____	*Too low*
10	$16(10)^2 = 16 \times $ _____ $ = $ _____	_____
_____	$16(__)^2 = 16 \times $ _____ $ = $ _____	_____
_____	_____	_____

Solution: The pebble will fall 576 feet in _____ seconds.

Look Back

What other strategy could you use to solve this problem?

_____

Mathematics Problem Solving Coach, Level G: Strategies and Applications

Practice Test Questions
Find Out • Plan • Solve • Look Back

1. Bernard has a square worktable with an area of 16 square feet. He says that the expression $s^2 + 2s + 1$ can be used to describe the area of a square whose sides are $(s + 1)$ units. What value of s will give the expression a value of 16? (Hint: Try the *make a table* strategy.)

   A 3
   B 4
   C 15
   D 17

2. Sally has a storage cube with a volume of 512 cubic inches. What is the length of an edge? (Hint: Try the *predict and test* strategy.)

Volume of a cube = edge³

   A 9 inches
   B 8 inches
   C 7 inches
   D 6 inches

3. Wilson drew this graph to show the solution to an equation.

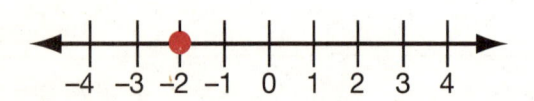

   ```
   ←――+――+――●――+――+――+――+――+――→
      -4  -3  -2  -1  0   1   2   3   4
   ```

   Which equation did he solve? (Hint: Try the *work backward* strategy.)

   A $n + 2 = 4$
   B $2 - n = 0$
   C $n + 3 = 1$
   D $n - 4 = 2$

4. The first step in Alvin's solution of an equation was $4n = 28$. Which could be the original equation?

   A $4n - 3 = 25$
   B $4n + 3 = 25$
   C $4n - 3 = 31$
   D $4n + 3 = 32$

5. Which equation does not have the same solution as $2x = 18$?

   A $4x = 36$
   B $6x = 54$
   C $8x = 96$
   D $20x = 180$

6. A community center rents its gym to local clubs. The fee for 3 hours is $275. This includes $50 for cleaning services. What is the hourly rental rate in dollars?

   Record your answer. Then fill in the bubbles. Be sure to use the correct place value.

⓪	⓪
①	①
②	②
③	③
④	④
⑤	⑤
⑥	⑥
⑦	⑦
⑧	⑧
⑨	⑨

Lesson 24: Expressions, One-Step, and Two-Step Equations

7. Nina has saved $20 toward the purchase of a $50 jacket. She figures she can save $6 per week. How many weeks does she need to save to have enough money to buy the jacket? (Hint: Try the *make a table* strategy.)

8. Emily earns $14 per hour at her weekend job. After one weekend, she put $100 of her pay in the bank and kept the remaining $54 for school supplies. How many hours did she work? Explain how you found your answer. (Hint: Try the *write an equation* strategy.)

_____

_____

_____

_____

9. A computer technician charges $38 for a house call and $29 per hour for labor. If the total bill comes to $125, how many hours did the repair take? (Hint: Try the *make a table* strategy.)

10. An amusement park charges $17.50 for an adult and $9.25 for a child. A group of 13 children and 5 adult counselors go to the park. How much do they pay to enter?

11. Earl buys his groceries at a store that gives 2.5% of all income to charity. This year, the store has an income of $2,400,000. How much will the store give to charity?

12. Vanna drives 39 miles to work. The trip averages 52 minutes. What is her average driving speed? Show how you got your answer.

Strategies: Diagram • Model • Organized List • Table/Graph • Pattern • Predict & Test
Logical Thinking • Work Backward • Simpler Problem • Formula or Equation

Part II: Applying the Strategies—Solving Problems with Algebra

25 Linear Equations

Linear equations have two variables and have graphs that are straight lines.

Example 1 Carey knows that these points are some of the solutions of a linear equation:

$B(0,1)$, $M(4,3)$, $R(6,4)$

Point K is also a solution of the equation. The x-coordinate of K is 2. What is the y-coordinate of K?

Find Out You know three solutions of a linear equation. You know the x-coordinate of another solution. You need to find the y-coordinate of that solution.

Plan You can use the *make a graph* strategy to solve this problem.

Solve Graph the points you know. Then draw a straight line through the points.

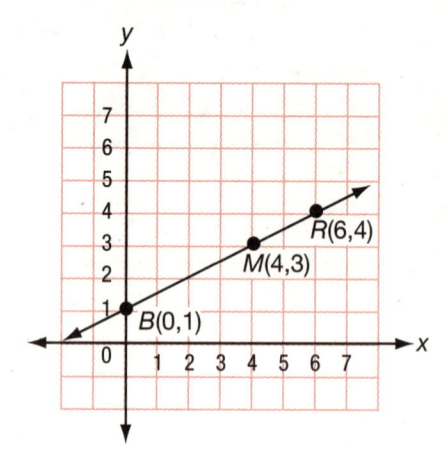

The x-coordinate of point K is 2. Draw a vertical line from 2 on the x-axis up to the graphed line. From the point of intersection, draw a horizontal line to the y-axis to find the y-coordinate.

Solution: The y-coordinate of point K is 2.

Look Back You can check your answer by looking for a pattern in the coordinates.

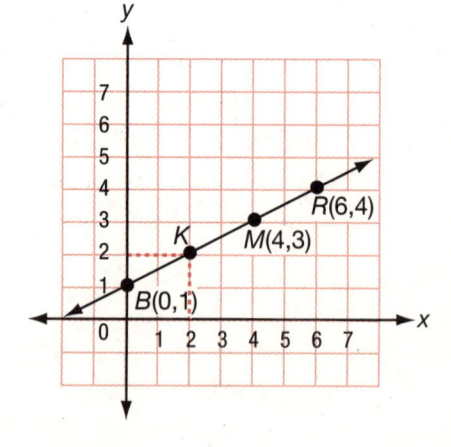

x: 0, 2, 4, 6   consecutive even numbers
y: 1, **2**, 3, 4   consecutive counting numbers ✔

Lesson 25: **Linear Equations**

Try It

Example 2

This function table shows the rise and run, in feet, of the first hill of a roller coaster.

Run (x)	Rise (y)
4	6
8	12
12	18
16	24

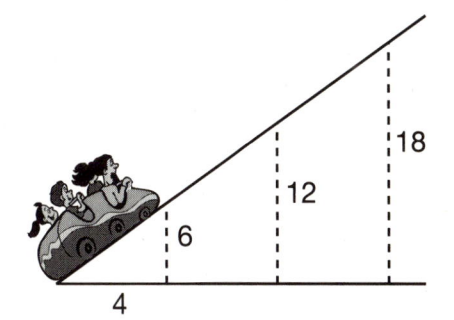

What equation can you use to describe this function?

Find Out You know corresponding rises and runs for the roller coaster hill. You need to find an equation for the function.

Plan You can use the *look for a pattern* strategy to solve this problem.

Solve Find the ratio of the rise to run for each pair of values.

$\frac{6}{4} =$ _____

$\frac{12}{8} =$ _____

$\frac{18}{12} =$ _____

$\frac{24}{16} =$ _____

The ratio of rise to run is always the same, so the function can be described using a linear equation.

What number times the input value equals the output value? _____

Solution: The equation is $y =$ _____ x.

Look Back How can you use the *work backward* strategy to check your work?

_____

_____

Mathematics Problem Solving Coach, Level G: Strategies and Applications

Practice Test Questions
Find Out • Plan • Solve • Look Back

1. Vicky drew a line parallel to this line. Which points could lie on her line? (Hint: Try the *make a table or graph* strategy.)

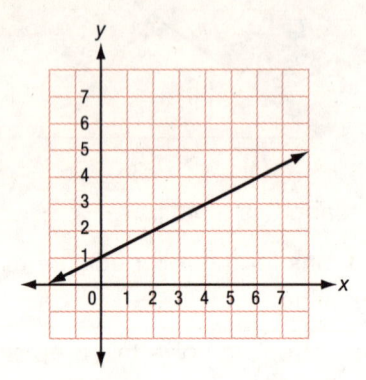

    A (1,0) and (2,2)   **C** (0,0) and (3,2)
    B (1,0) and (3,1)   **D** (0,2) and (3,5)

2. Evan located these points on the same line.

    P(1,3), Q(2,4), R(3,5), S(4,6)

Which is another point on the line? (Hint: Try the *look for a pattern* strategy.)

    A W(11,9)   **C** X(8,12)
    B V(9,11)   **D** Y(6,10)

3. Geri draws four lines. Which pair of points are on a line with a negative slope? (Hint: Try the *predict and test* strategy.)

    A F(3,6) and G(1,4)
    B H(5,8) and J(3,2)
    C P(4,9) and Q(3,4)
    D Not here

4. Maurice drew this graph of an equation. Which equation could it be?

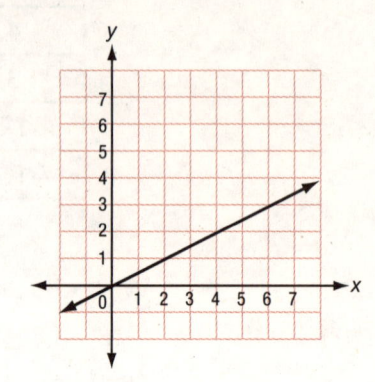

    A $y = x + 2$ **C** $y = \frac{1}{2}x$
    B $y = x + \frac{1}{2}$ **D** $y = 2x$

5. Which of the following ordered pairs is a solution of $y = 2x - 3$?

    A (1,−1)   **C** (−1,5)
    B (1,1)   **D** (2,7)

6. Dixie drew a line that passes through all these points.

    (12,8), (16,10), (24,___), (30,17)

What is the missing *y*-coordinate?

Record your answer. Then fill in the bubbles. Be sure to use the correct place value.

Lesson 25: Linear Equations

7. The graph of a slide passes through points with coordinates (4,7) and (1,3). What is the slope of the slide? (Hint: Try the *write an equation* strategy.)

8. Sherri bought some CDs at $12 each and some tapes at $9 each. She bought more CDs than tapes. She spent a total of $54. What linear equation models her purchases? How many CDs did she buy? (Hint: Try the *predict and test* strategy.)

9. Damien drew the graph of these two equations on the same coordinate grid.

$y = x + 1$ and $y = 2x$

Draw a graph of the equations and name the coordinates of the point where the lines with these equations intersect. (Hint: Try the *make a table or graph* strategy.)

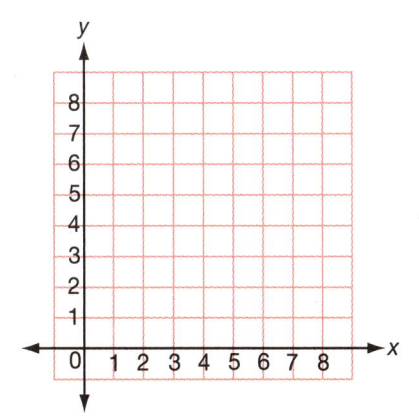

10. A skateboard ramp has a slope of $\frac{1}{12}$. The top of the ramp reaches a platform that is 3 feet high. What is the distance from the bottom of the ramp to the base of the platform? Explain how you found your answer.

_____
_____
_____
_____

11. A line has a slope of $\frac{2}{3}$ and passes through the point with coordinates (5,4). On the same line, find the point having positive integer coordinates that are less than 5 and 4. Explain your answer.

_____
_____
_____
_____

12. A line has the equation $y = 3x + 2$. Find an ordered-pair solution in which the y-coordinate is 10 more than the x-coordinate.

Strategies: Diagram • Model • Organized List • Table/Graph • Pattern • Predict & Test
Logical Thinking • Work Backward • Simpler Problem • Formula or Equation

Part II: Applying the Strategies—Solving Problems with Data and Probability

26 Measures of Central Tendency

You can use mean, median, mode, range, and quartiles to describe sets of data. Each gives different information about the data.

Example 1   Charlie asked his classmates how much television they watched the night before. These are his results.

How much television did you watch last night?

Time (in minutes)	0	30	60	90	120	150	180
Number of Students	1	0	0	4	5	3	7

What are the mean, median, mode, and the range of the times?

Find Out   You want to find the mean, median, mode, and range of the amount of television students watched.

Plan   You can use the *write an equation* strategy to solve the problem.

Solve   **Find the mean.** The mean is the sum of all the items divided by the number of items. There are 20 items, so divide by 20.

$(1 \times 0) + (4 \times 90) + (5 \times 120) + (3 \times 150) + (7 \times 180) = 2{,}670$
$2{,}670 \div 20 = 133.5$

Find the median. The median is the middle item when the items are listed from least to greatest. There are 20 items, so there are two middle terms. The tenth item is 120. The eleventh item is 150.

$(120 + 150) \div 2 = 135$ *Find the mean of the middle items.*

Find the mode. The mode is the item that occurs most often. The greatest number of students watched for 180 minutes.

Find the range. The range is the difference between the greatest and least items.

$180 - 0 = 180$

Solution: The mean is 133.5 minutes. The median is 135 minutes. The mode is 180 minutes. The range is 180 minutes.

Look Back   Use the definitions of the terms and double-check your math to make sure that your answers are correct. ✔

Lesson 26: **Measures of Central Tendency**

Try It

Example 2 Mr. Gomez is taking students with averages in the upper quartile to the movies and then to eat pizza. These are the averages of the students in his class. Who will go out with Mr. Gomez?

Math Averages			
Anton	95	Jerry	94
Arlene	85	Jim	88
Bijan	87	Myra	86
Constance	67	Pablo	80
David	75	Peter	81
Dena	73	Raji	83
Diedre	82	Sachiko	93
Desmond	92	Susan	91
Helene	71	Toby	88
Hicham	90	Zachary	82

Find Out You know all the averages. You also know that quartiles divide the items into four equal parts. The lower quartile is the median of the lower half. The upper quartile is the median of the upper half. You want to find the scores in the fourth quartile.

Plan You can use the *make an organized list* strategy to solve the problem.

Solve Write all the scores in order, from lowest to highest.

67, 71, 73, 75, 80, _____

Count the number of scores. _____

The upper quartile is the upper $\frac{1}{4}$ of the scores.

$\frac{1}{4}$ × number of scores = $\frac{1}{4}$ × _____ = _____

Ring the scores in the upper quartile on your list.

Solution: The students who will go out with Mr. Gomez are _____

_____

Look Back How could you check by solving this problem another way?

Mathematics Problem Solving Coach, Level G: Strategies and Applications

Practice Test Questions
Find Out • Plan • Solve • Look Back

Use the table for problems 1 and 2.

Math Test Scores

Hope	Kenji	Lily	Sean
95	85	90	88
87	87	82	85
90	84	80	94
86	90	85	90
92	89	78	97
90	91	81	85

1. Who has the highest mean score? (Hint: Try the *write an equation* strategy.)

   A Hope
   B Kenji
   C Lily
   D Sean

2. What is Lily's median score? (Hint: Try the *make an organized list* strategy.)

   A 81
   B 81.5
   C 82
   D 82.7

3. Randy's math test scores are 98, 87, 88, 90, 92, 95, 99, and 89. How many of her test scores are greater than her mean score? (Hint: Try the *solve a simpler problem* strategy.)

   A 1   C 3
   B 2   D 4

4. Gabe, Brady, Harold, and Mel collect baseball cards. Gabe has 32, Brady has 48, Harold has 29, and Mel has 43. Mel gives Gabe 8 cards. Which statement is true?

   A The upper quartile changes.
   B The median changes.
   C The mean changes.
   D Not here

5. Lena recorded the number of sit-ups she did each day for a week.

   13, 15, 16, 18, 13, 14, 16

   Which number is not a measure of central tendency?

   A 13   C 15
   B 14   D 16

6. This list shows the distance run in miles by each member of the track team.

   5.80, 6.20, 6.15, 6.05, 5.95, 6.15

   What is the mean distance run?

   Record your answer. Then fill in the bubbles. Be sure to use the correct place value.

Lesson 26: **Measures of Central Tendency**

7. The table shows the life expectancies in some countries around the world.

Country	Life Expectancy (in years)
Argentina	66.7
Australia	73.2
China	62.3
Cuba	68.4
Greece	72.5
Italy	72.7
Japan	74.5
Spain	72.8
Sweden	73.0
Switzerland	72.5
United States	70.0

What is the range between the countries with the longest and shortest life expectancies? Which countries are they? (Hint: Try the *solve a simpler problem* strategy.)

8. Tate's 5 science test scores have a range of 8 and a median of 86. His first 4 test scores are 88, 82, 86, and 83. What is his fifth test score? (Hint: Try the *predict and test* strategy.)

9. For the following data, which two measures of central tendency are the same—mean, median, mode, range? (Hint: Try the *use logical thinking* strategy.)

19, 12, 15, 12, 22, 10, 12, 20, 21, 20

10. Mrs. Emerson made a histogram of the class test scores.

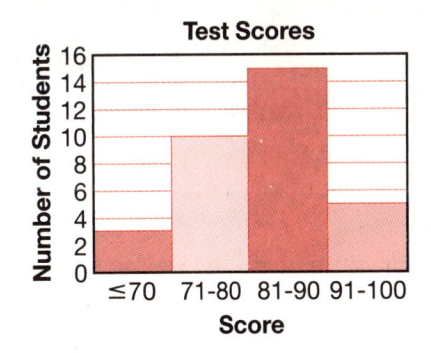

Where will the median be? Explain how you know.

_____
_____
_____
_____

The tally table shows the results of a student survey. Use this table for problems 11 and 12.

Height (in inches)	Number of Students
43	II
44	I
45	ℍℍ I
46	ℍℍ ℍℍ
47	ℍℍ III
48	III

11. What is the mean height of students in the survey?

12. What is the lower quartile?

Strategies: Diagram • Model • Organized List • Table/Graph • Pattern • Predict & Test
Logical Thinking • Work Backward • Simpler Problem • Formula or Equation

Part II: Applying the Strategies—Solving Problems with Data and Probability

27 Combinations and Permutations

An arrangement where the order of items is important is a *permutation*. An arrangement where the order does not matter is a *combination*. You can use a tree diagram, use the Fundamental Counting Principle, or make a list to help find the number of possible outcomes.

Example 1

A computer store sells 12 different monitors, 3 different keyboards, 5 different printers, and 7 different computer models. Clarissa wants to buy a new monitor and a new printer. From how many combinations can she choose?

Find Out

You know the number of different monitors and printers. You want to find out how many different combinations there are.

Plan

You can *make an organized list* to solve the problem.

Solve

Use the numbers 1 to 12 to stand for the different monitors.
Use the letters A to E to stand for the different printers.

1A	2A	3A	4A	5A	6A	7A	8A	9A	10A	11A	12A
1B	2B	3B	4B	5B	6B	7B	8B	9B	10B	11B	12B
1C	2C	3C	4C	5C	6C	7C	8C	9C	10C	11C	12C
1D	2D	3D	4D	5D	6D	7D	8D	9D	10D	11D	12D
1E	2E	3E	4E	5E	6E	7E	8E	9E	10E	11E	12E

There are 60 different combinations.

Solution: Clarissa can choose from 60 monitor and printer combinations.

Look Back

You can *write an equation* to check your answer. Use the Fundamental Counting Principle to find the number of combinations.
There are 12 choices of monitor.
There are 5 choices of printer.

number of choices = number of monitors × number of printers
$n = 12 \times 5$
$n = 60$ ✔

Lesson 27: **Combinations and Permutations**

Try It

Example 2 Jack, Yoshi, Abby, Rosalita, and Bruce are running in the election for soccer team managers. There will be two managers. How many outcomes can the election have?

Find Out You want to find the number of ways to choose two team managers.

Plan You can use the *draw a diagram* strategy to solve the problem.

Solve To find the number of outcomes, find the number of permutations possible and subtract the number of duplicates.

Make a tree diagram to show all the possible pairs.

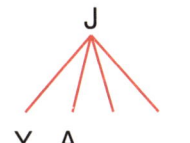

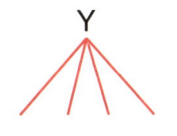

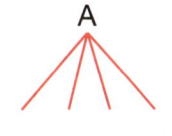

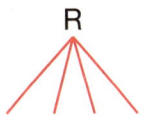

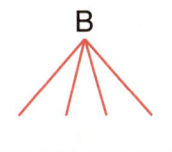

```
      J           Y           A           R           B
     /|\         /|\         /|\         /|\         /|\
    / | \       / | \       / | \       / | \       / | \
   Y  A __    __ __ __    __ __ __    __ __ __    __ __ __
```

- Ring pairs that are duplicates.
- Count the remaining pairs.

Solution: The election can have _____ outcomes.

Look Back How could you check your answer by solving this problem another way? Show your work.

_____
_____

Practice Test Questions
Find Out • Plan • Solve • Look Back

1. In how many different ways can Dennis, Ervin, Roxie, Luci, Elaine, and Phillip pair up to play table tennis? (Hint: Try the *make an organized list* strategy.)

   A 12
   B 15
   C 360
   D 720

2. A pizza restaurant offers 6 different toppings: mushrooms, olives, sausage, meatballs, pepperoni, and peppers. How many different pizzas are possible if you order two different toppings? (Hint: Try the *draw a diagram* strategy.)

   A 720
   B 30
   C 12
   D Not here

3. Members of the track club wear two pairs of socks when they practice. Socks come in white, gray, black, red, and blue. How many different combinations of socks can they have? (Hint: Try the *write an equation* strategy.)

   A 25
   B 20
   C 10
   D 5

4. Zander and Violet are playing a board game. For each turn, the player rolls two 1–6 number cubes and spins the spinner.

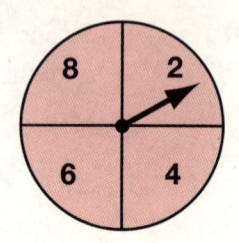

   How many possible outcomes are there?

   A 16
   B 24
   C 36
   D 144

5. You rent 5 videos for a rainy weekend but have time to watch only 3. How many different combinations of videos can you watch?

   Record your answer. Then fill in the bubbles. Be sure to use the correct place value.

Lesson 27: **Combinations and Permutations**

Use this information for problems 6 and 7.

The Ice Cream Palace
12 Flavors
6 Toppings
4 Cones
3 Whipped Cream Flavors

6. Sandy wants one scoop of ice cream in a cone with a topping and whipped cream. How many choices does she have? (Hint: Try the *solve a simpler problem* strategy.)

7. Mark wants 2 different scoops in a cone. He wants one of the scoops to be mint chocolate chip. How many choices does he have? (Hint: Try the *use logical thinking* strategy.)

8. How many 4-digit numbers can you make from the digits 2, 4, 6, and 8? What fraction of the numbers will have 6 in the hundreds place? (Hint: Try the *make an organized list* strategy.)

9. Four students are running for the Student Council. The student who receives the most votes will become President. The student with the second-highest vote total will become Vice-President. How many possible combinations are there for these two positions?

10. At the grocery store, Myron decides whether to use the self-checkout or the full-service line. He decides whether to use cash, check, credit card, or debit card. He also decides whether to use plastic or paper bags. How many possible outcomes are there?

11. A committee is selecting 2 sculptures for the new town library. Altogether, 8 sculptures have been submitted to the committee for judging. How many possible combinations are there? How many possible permutations are there? Explain why these are different.

_____
_____
_____
_____

Strategies: Diagram • Model • Organized List • Table/Graph • Pattern • Predict & Test
Logical Thinking • Work Backward • Simpler Problem • Formula or Equation

Part II: Applying the Strategies—Solving Problems with Data and Probability

28 Experimental and Theoretical Probability

The *probability* of an event happening is a number between 0 and 1. Another way of expressing probability is to use odds. The *odds* of an event happening are the ratio of favorable outcomes to unfavorable outcomes.

Example 1

Find Out

Denise tosses three 1–6 number cubes. What is the probability that the sum of the numbers is 18?

You want to find the probability of tossing three number cubes and getting a sum of 18.

Plan

This is a multistep problem. First, you can use the *make an organized list* and *find a pattern* strategies to find the number of outcomes. Then you can find the outcomes where the sum is 18. Finally, you can find the probability of such an outcome.

Solve

Make a list of all the outcomes when the number on the first cube is 1.

1 1 1	1 2 1	1 3 1	1 4 1	1 5 1	1 6 1
1 1 2	1 2 2	1 3 2	1 4 2	1 5 2	1 6 2
1 1 3	1 2 3	1 3 3	1 4 3	1 5 3	1 6 3
1 1 4	1 2 4	1 3 4	1 4 4	1 5 4	1 6 4
1 1 5	1 2 5	1 3 5	1 4 5	1 5 5	1 6 5
1 1 6	1 2 6	1 3 6	1 4 6	1 5 6	1 6 6

There are 36 outcomes when the first cube is 1. So, there are 36 outcomes when the first cube is 2. The same is true when the first cube is 3, 4, 5, and 6.

The total number of outcomes = 6 × 36 = 216.

There is only one favorable outcome with a sum of 18. All three number cubes must show 6.

Solution: The probability that the sum of the numbers is 18 is $\frac{1}{216}$.

Look Back

You can use the Fundamental Counting Principle to find the total number of outcomes.

Each number cube has 6 outcomes. The total number of outcomes for three number cubes is 6 × 6 × 6, or 216. The probability of tossing a sum of 18 with three number cubes is $\frac{1}{216}$. ✓

Lesson 28: Experimental and Theoretical Probability

Try It

Example 2 A teacher writes the name of each state on an index card and puts all the cards in a bag. Students must write reports on the two states they pick. Juliet is first to pick. She hopes to pick Hawaii and Alaska. What are the odds that she will do so?

Find Out You want to find the odds that Juliet will pick Hawaii and Alaska.

Plan You can use the *write an equation* strategy to solve the problem.

Solve On Juliet's first pick, there are _____ possible outcomes.

Of these, _____ are favorable.

The probability of a favorable outcome on the first pick is _____.

If Juliet picks one of her choices on the first pick, there is _____ favorable outcome on the second pick.

On the second pick there are _____ possible outcomes.

The probability of a favorable outcome on the second pick is _____.

The probability of picking both states is _____ × _____ = _____.

The probability of picking both states is = _____.

Write the ratio of the probability of picking both states to the probability of not picking both states. Simplify if possible.

probability of picking both states ⟶

probability of **not** picking both states ⟶

$$\frac{\Box}{\Box} = \frac{\Box}{\Box}$$

Solution: The odds that that Juliet will pick her two choices are _____ to _____.

Look Back How can you check your answer?

_____

Mathematics Problem Solving Coach, Level G: Strategies and Applications

Practice Test Questions
Find Out • Plan • Solve • Look Back

1. If you toss 2 quarters, there is a 25% chance that the result will be 2 heads. What are the odds that you will not get 2 heads? (Hint: Try the *make an organized list* strategy.)

    A 2 to 1
    B 3 to 1
    C 1 to 3
    D 1 to 4

2. Matt has 2 red counters, 4 yellow counters, and 6 blue counters in a bag. He pulls out 2 counters. What is the probability that he pulls out a red and a yellow counter? (Hint: Try the *write an equation* strategy.)

    A $\frac{1}{6}$ C $\frac{1}{18}$
    B $\frac{4}{11}$ D $\frac{2}{33}$

3. Shelly spins these two spinners.

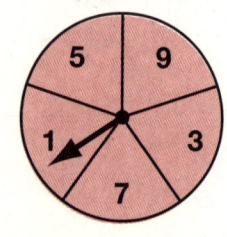

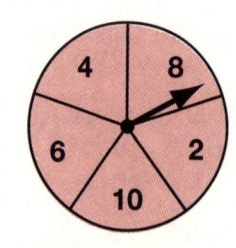

    What is the probability that she spins a sum greater than 13? (Hint: Try the *look for a pattern* strategy.)

    A $\frac{6}{25}$ C $\frac{1}{2}$
    B $\frac{2}{5}$ D $\frac{3}{5}$

4. Jimmy has a stack of cards. He has 4 kings, 4 queens, and 4 jacks. He lays them face down on the table. He picks a card. Which statement is **not** true?

    A The probability of picking a queen is $\frac{1}{3}$.
    B The odds of picking a jack are 1 to 2.
    C The probability of not picking a king is $\frac{1}{3}$.
    D The odds of picking a queen or a jack are 2 to 1.

5. Ms. Burton is holding some coins in her hand. She has 10 quarters and some nickels. She says that the odds of picking a nickel are 1 to 2. How many nickels is she holding?

    Record your answer. Then fill in the bubbles. Be sure to use the correct place value.

Lesson 28: **Experimental and Theoretical Probability**

6. In his sock drawer, Alfredo has 6 black socks, 10 white socks, and 4 blue socks. What is the probability that Alfredo will reach into the drawer without looking and choose a pair of black socks? (Hint: Try the *write an equation* strategy.)

7. The odds of thunderstorms today are 2 to 3. What is the probability that there will be no thunderstorms today? (Hint: Try the *work backward* strategy.)

8. Rodney conducted a survey in his class by asking students how many CDs they each have. The stem-and-leaf plot shows the results of the survey.

Student CDs

Stem	Leaves
1	2 5 8 9
2	0 0 4 4 5 7 8 8
3	0 1 3 5 5 5 5 8

Key: 1|2 means 12

What are the odds that a student selected randomly will have 35 CDs? (Hint: Try the *solve a simpler problem* strategy.)

Annika has a bag containing 15 purple marbles, 12 white marbles, and 9 black marbles. Use this information for problems 9 and 10.

9. What is the probability that she will choose a black marble, replace it, and then choose a white marble?

10. What are the odds that Annika will choose a purple marble?

11. Logan has a stack of 36 numbered cards. He says that the odds of picking a 4 are 1 to 8 and the probability of not picking a 7 is $\frac{13}{18}$. How many 4s and 7s does Logan have?

Strategies: Diagram • Model • Organized List • Table/Graph • Pattern • Predict & Test
Logical Thinking • Work Backward • Simpler Problem • Formula or Equation

Math Level G Practice Test
Answer Sheet

Name: _____

1. A B C D
2. A B C D
3. A B C D
4. A B C D
5. A B C D
6. A B C D
7. A B C D
8. A B C D
9. A B C D
10. A B C D
11. A B C D
12. A B C D
13. A B C D
14. A B C D
15. A B C D
16. A B C D
17. A B C D
18. A B C D
19. A B C D
20. A B C D
21. A B C D
22. A B C D
23. A B C D
24. A B C D
25. A B C D
26. A B C D
27. A B C D
28. A B C D
29. A B C D
30. A B C D

31. A B C D
32. A B C D
33. A B C D
34. A B C D
35. A B C D
36. A B C D
37. A B C D
38. A B C D
39. _____
40. _____
41. _____
42. [grid-in 0-9]
43. _____
44. _____
45. [grid-in 0-9]

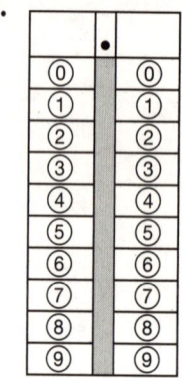

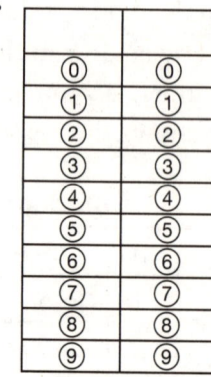

46. _____
47. [grid-in 0-9]
48. _____
49. _____
50. _____
51. [grid-in 0-9]
52. _____
53. _____
54. _____
55. _____
56. _____

Math Level G Practice Test
Test-Taking Tips

When you take the Practice Test, read each question carefully. Answer all the questions. If you're not sure of an answer, make the best guess you can. Always ask yourself if your answer makes sense. If you have time, go back and check your answers before you hand in the test.

Tips for Multiple-Choice Questions
1. Read all the choices before you choose an answer.
2. If you're not sure of the right answer, cross out the wrong answers. Then make the best guess you can.
3. Mark the correct answer on the answer sheet or circle the letter of the correct answer on the test page. If you change an answer, neatly erase the answer you don't want.

Tips for Using the Bubble Grid or Writing an Answer
1. After you decide on the answer, check it. Usually, you can start with your answer and work backward. If you end up with the information given in the problem, your answer should be correct.
2. For the bubble grid: Write the answer in the top row. Then fill in the bubbles that match. Be sure to use the correct place value.
3. When you write out an answer, write clearly. If you change the answer, neatly erase or cross out the answer you don't want.

How to Record your Answers
Sample question:

1. There are 12 inches in 1 foot. How many inches are there in 3 feet?

Working on the test sheet

    A 12 inches
    B 1 yard
  (C) 36 inches
    D 12 feet

Working on the answer sheet

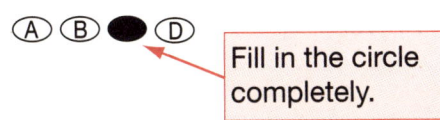

Ⓐ Ⓑ ● Ⓓ   Fill in the circle completely.

Writing out the answer

36 inches
$12 \times 3 = 36$

Sometimes you have to show your work.

Using the bubble grid

3	6
⓪	⓪
①	①
②	②
●	③
④	④
⑤	⑤
⑥	●
⑦	⑦
⑧	⑧
⑨	⑨

Math Word Problem Level G
Practice Test

1. Ivana sees the following road sign.

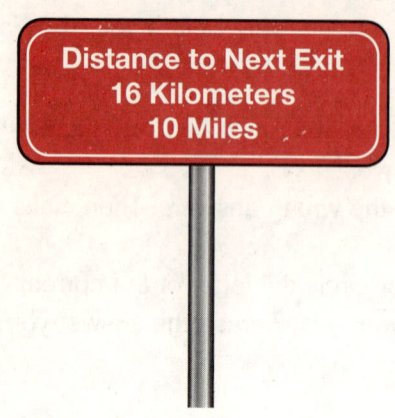

   Distance to Next Exit
   16 Kilometers
   10 Miles

   Ivana is 25 miles from her hotel. About how many kilometers is that?

   A 16 kilometers
   B 40 kilometers
   C 400 kilometers
   D Not here

2. Finn tells Christina that the sum of three consecutive integers is always divisible by both 2 and 3. Which set of consecutive integers can Christina use to show to Finn that he is incorrect?

   A 3, 4, and 5
   B 5, 6, and 7
   C 6, 7, and 8
   D 7, 8, and 9

3. Alexis does volunteer work at a local hospital. During her first week, she works for 12 hours. After that, she works for 8 hours per week. How many weeks will it take her to work a total of 100 hours?

   A 8 weeks
   B 11 weeks
   C 12 weeks
   D 15 weeks

4. Triangle ABC is a right triangle with an area of 10 square units. Which of the following represents the coordinates of a triangle congruent to triangle ABC?

   A (0,0), (4,0), and (0,5)
   B (0,0), (3,0), and (0,7)
   C (0,0), (5,0), and (0,2)
   D (0,0), (6,0), and (0,4)

5. Carl's class holds an auction to raise money for an end-of-the-year party. Carl guesses that his class will raise $150, but they actually raise 18% more than that. How much money does his class raise?

   A $168
   B $177
   C $183
   D $185

6. Pedro rents 4 videotapes and 2 DVDs. Which of the following, by itself, does **not** give you enough information to determine the total amount that Pedro spends on his rentals?

A It costs $4 to rent a videotape and $1\frac{1}{2}$ times as much to rent a DVD.

B It costs $6 to rent a DVD and $2 less to rent a videotape.

C It costs $4 to rent a videotape and $2 more than that to rent a DVD.

D It costs $10 to rent one videotape and one DVD.

7. Della needs to buy T-shirts to bring to summer camp. If she buys 12 T-shirts at the sale price, how much money does she save?

Regular Price $10 each

Sale Price 3 for $25

A $15
B $20
C $25
D $30

8. Miles goes to an Internet café with his friends. It costs $5.50 an hour to rent a computer and $1.50 for a soda. Miles starts out with $25. Which expression shows how much money he will have left after he rents a computer for h hours and buys 2 sodas?

A $25.00 − 5.50h$ − $3.00

B $25.00 − 3.00h$ − $5.50

C $25.00 − 5.50h$ − $1.50

D $25.00 + 5.50h$ + $3.00

9. Point $A(3,1)$ is rotated a quarter turn clockwise and then reflected across the y-axis. In which quadrant is the image of the point?

A Quadrant I
B Quadrant II
C Quadrant III
D Quadrant III

10. If the lengths of all four sides of a square are tripled, the perimeter of the new square will be 48 inches. What is the length of a side of the original square?

perimeter = 4 × length

A 3 inches
B 6 inches
C 16 inches
D Not here

11. Pat rides his skateboard over the ramp shown below.

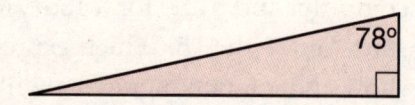

What is the measure of the angle that the ramp makes with the ground?

A 12°
B 14°
C 18°
D 22°

12. In May 2003, Robert Cheruiyot won the Boston Marathon by running the 26.2-mile race in 2 hours, 10 minutes, 11 seconds. Which of the following best approximates his average speed?

A 4.5 minutes per mile
B 5 minutes per mile
C 5.5 minutes per mile
D 6 minutes per mile

13. Kathryn is in charge of music for the class dance. She decides to bring one CD by each of her three favorite groups. If she owns 3 CDs by one of these groups, 4 CDs by the second group, and 2 CDs by the third group, in how many different ways can she make her selection?

A 9
B 11
C 14
D Not here

14. Joanne plans to fence in her 9-meter by 12-meter rectangular garden. If she puts a fence post every 3 meters, how many posts does she need?

perimeter = 2 × (length × width)

A 20
B 18
C 16
D 14

15. The two polygons shown below are similar, but the length of one of the sides of polygon ABCDE has been labeled incorrectly.

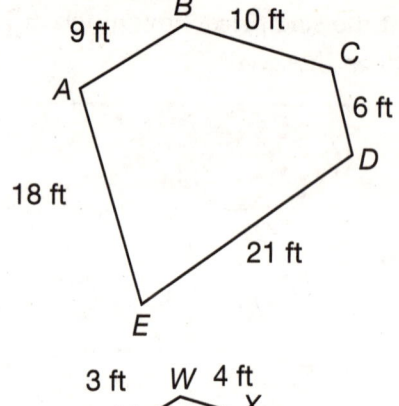

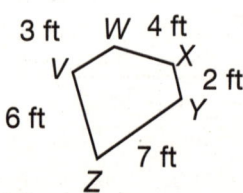

If side AB corresponds to side VW, and ∠A corresponds to ∠V, which side of polygon ABCDE has the incorrect length?

A side BC
B side CD
C side DE
D side AE

16. A music store is having a raffle for a portable CD player. There are 400 raffle tickets in all. The winner will be chosen at random. Andrew has 5 tickets, and his brother Brendan has 3 tickets. What is the probability that one of the brothers will win?

  A $\frac{1}{50}$

  B $\frac{1}{80}$

  C $\frac{1}{400}$

  D $\frac{3}{400}$

17. What is the ratio of the volume of a cube with sides that are $\frac{1}{2}$ inch long to the volume of a cube with sides that are 1 inch long?

  A $\frac{1}{16}$

  B $\frac{1}{8}$

  C $\frac{1}{4}$

  D $\frac{1}{2}$

18. A line goes through the point (0,0) and has a slope of 3. Which of the following points is on the line?

  A (1,3)

  B (3,3)

  C (3,1)

  D Not here

19. Alvaro drives directly from Maplewood to Somerville and then to Union.

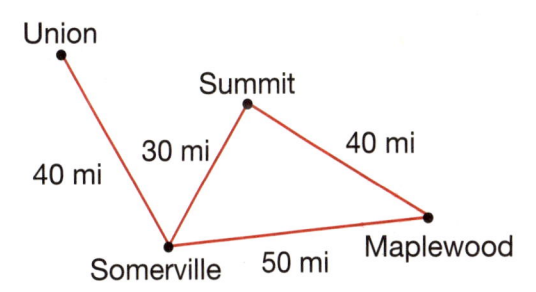

If it takes him 2 hours to make the trip, what is his average speed?

  A 45 miles per hour

  B 47 miles per hour

  C 50 miles per hour

  D 55 miles per hour

20. Which of these numbers has a square that is less than the number itself?

  A 3

  B 1

  C $\frac{1}{2}$

  D -2

21. A rectangular storage shed has an area of 108 square feet and a perimeter of 42 feet. What is the length of the longer side?

> Perimeter = 2 × (length + width)
> Area = length × width

  A 9 feet

  B $10\frac{1}{2}$ feet

  C 12 feet

  D 21 feet

22. Jon plays football for his school team. At the start of a game, he runs the ball three times in a row, gaining a total of 9 yards. On the first play, he gains 17 yards, and on the second play he loses 4 yards. What happens on the third play?

A He gains 4 yards.
B He loses 4 yards.
C He gains 21 yards.
D He loses 21 yards.

23. Which number line best shows the location of $\sqrt{17} + \sqrt{26}$?

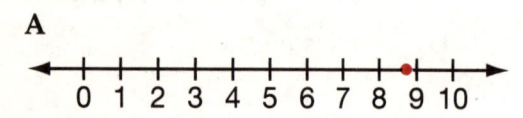

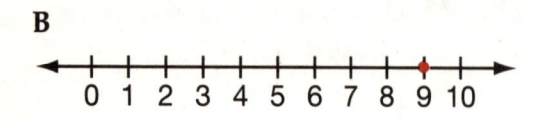

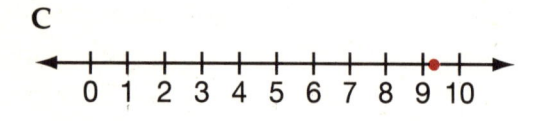

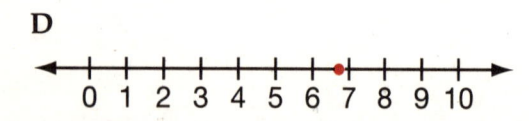

A
B
C
D

24. The mean of three numbers is 19 and the range is 8. What are the three numbers?

A 15, 20, 25
B 20, 26, 28
C 15, 19, 23
D 19, 23, 27

25. In the figure below, line segment HI intersects line GJ at point H. The measure of ∠GHI is 9 times the measure of ∠IHJ. What is the measure of ∠IHJ?

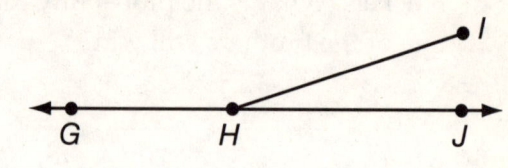

A 15°
B 18°
C 20°
D Not here

26. A shipment of 50 schoolbooks is delivered in one large box. The box contains 30 two-pound books. The rest of the books weigh 4 pounds. The shipping charge is $0.25 per pound. What is the total weight of the books?

Which of the facts is **not** needed to solve the problem?

A The shipment contains 50 books.
B The shipment contains 30 two-pound books.
C All of the books weigh either 2 pounds or 4 pounds.
D The shipping charge is $0.25 per pound.

Practice Test

27. One member of the Student Council will be chosen at random to participate in a special citywide meeting. The table below shows the number of students in the Student Council.

Grade	7	8	9
Boys	2	8	11
Girls	5	4	15

What is the probability that the student selected at random will be a girl from the seventh grade? Express your answer in simplest form.

A $\frac{1}{9}$ 　　　　C $\frac{5}{24}$

B $\frac{7}{45}$ 　　　　D $\frac{24}{45}$

28. A rectangle has vertex coordinates of (3,2), (3,5), (7,2), and (7,5). After a translation, the images of the first three vertices are (8,1), (8,4), and (12,1). What are the coordinates of the fourth vertex?

A (12,6)
B (7,4)
C (12,4)
D Not here

29. If you flip a penny and roll a number cube numbered 1 through 6, what is the probability that you will land on tails and roll a number greater than 4?

A $\frac{1}{12}$ 　　　　C $\frac{1}{4}$

B $\frac{1}{6}$ 　　　　D $\frac{1}{2}$

30. Byron is building a model railroad village on this circular board.

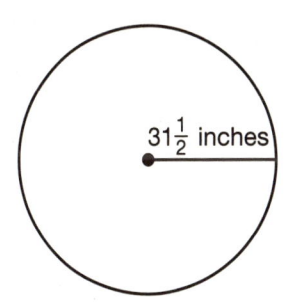

$31\frac{1}{2}$ inches

He plans to cover the entire board with special paper that looks like grass. Which of the following represents the smallest sheet of paper that is big enough to cover the entire board?

Area = π × radius²

A 4 feet by 4 feet
B 6 feet by 6 feet
C 8 feet by 6 feet
D 8 feet by 8 feet

31. It takes Tiffany 3 hours, 23 minutes, 48 seconds to finish her homework. Which expression represents the total amount of time in seconds that it takes?

A (3 × 3,600) + (23 × 60) + 48
B (3 × 24) + (23 × 60) + 48
C (48 × 3,600) + (23 × 60) + 3
D (3 × 60) + (23 × 60) + 48

32. In which group are all the numbers between $\sqrt{15}$ and 6?

A 5.2, $\sqrt{35}$, $\frac{33}{8}$ 　　C 4.9, $\frac{55}{9}$, $\sqrt{29}$

B 4.7, $\sqrt{37}$, $\frac{17}{3}$ 　　D 3.6, $\frac{28}{5}$, 21

33. Joel baby-sat on Saturday. This week, he spent $\frac{1}{4}$ of his earnings on a movie and $\frac{1}{6}$ on food. He has $14 left. How much did he earn baby-sitting?

   A $24
   B $26
   C $28
   D $30

34. A, B, and C, are three different points on a straight line. B is between A and C. Which of the following **cannot** be true?

   A Segments AB and BC have the same length.
   B Segment AC is longer than segment AB.
   C Segment BC is shorter than segment AB.
   D Segment AC is shorter than segment BC.

35. The greatest prime factor of a positive integer is 7. Which of the following could **not** be the positive integer?

   A 35
   B 42
   C 63
   D 77

36. What is the missing number in the sequence?

   $8, -4, 2, -1, \frac{1}{2}, \underline{\ ?\ }, \frac{1}{8}$

   A $\frac{1}{3}$
   B $-\frac{1}{3}$
   C $\frac{1}{4}$
   D $-\frac{1}{4}$

37. Holly is choosing a wristwatch for her birthday. The style that she likes comes in 4 different colors. There are 5 different choices of watchband. From how many different types of watches can she choose?

   A 9
   B 13
   C 24
   D Not here

38. Cary recorded the temperature every day at noon. The graph shows the information she recorded.

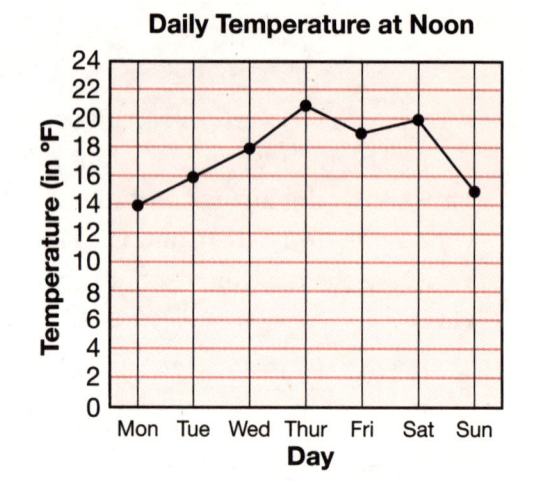

   Daily Temperature at Noon

   Which measure of central tendency does 18° represent?

   A upper quartile
   B mean
   C mode
   D median

Practice Test

39. As a part of an Earth Day project on recycling, Hank found the following information.

Trash in U.S. Landfills

Type of Trash	Percent
Metal	8%
Plastic	24%
Food and Yard Waste	11%
Rubber and Leather	6%
Other Trash	21%
Paper	30%

He displays the information in this circle graph.

Trash in U.S. Landfills

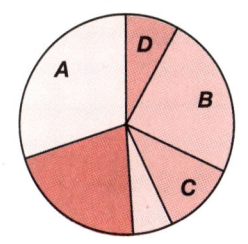

Which section of the graph represents the percent of plastic in landfills? Explain your answer.

_____

_____

_____

40. The radius of the planet Mercury is 1,516 miles. What is the circumference of Mercury to the nearest whole number? Use 3.14 for π.

Circumference = 2 × π × radius

41. Rick and Ben opened a new box of cereal and each boy ate a bowl of cereal. After this, the box contained cereal to the height shown below.

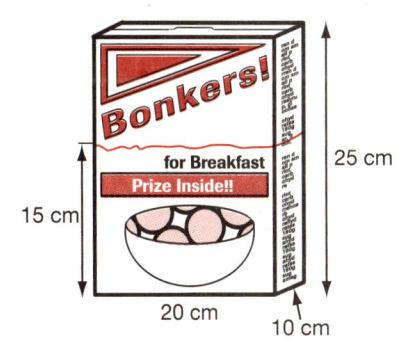

25 cm
15 cm
20 cm
10 cm

What percent of the cereal in the box did the boys eat? Explain your answer.

_____

_____

_____

42. Becky is on the track team. On Monday, she warms up at practice by running $\frac{3}{8}$ mile. If this represents 15% of the total amount that she runs during the practice, how many miles does she run in all?

Record your answer. Then fill in the bubbles. Be sure to use the correct place value.

109

43. Gus polls all of the students in his homeroom to find out their favorite sports. Gus allows each student to select only one sport. He summarizes the results in the graph below.

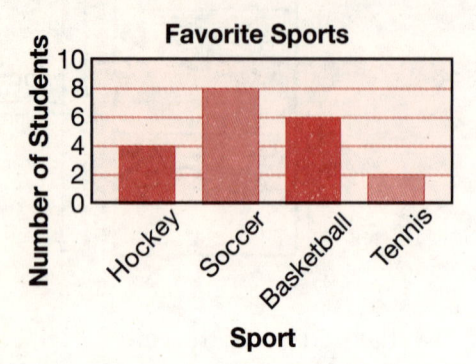

Favorite Sports
(Bar graph: Hockey = 4, Soccer = 8, Basketball = 6, Tennis = 2)

Using this data, draw a circle graph showing the **percent** of the homeroom that favors each sport.

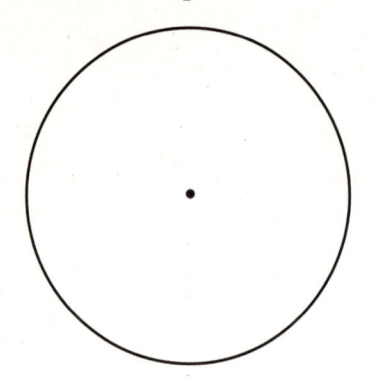

44. Josh coaches a roller hockey team with 6 skaters: Albert, Brian, Colin, Don, Erik, and Francis. Since only four skaters can play at a time, Josh needs to decide which two players will sit out at the start of the game. In how many different ways can he make this decision?

45. When Rachel uses her cell phone during the day, she pays 20¢ for the first minute and 7¢ for each minute after that. At night, she pays a fixed rate of 5¢ per minute. If Rachel plans to make a 9-minute call to her uncle, how many cents would she save by making the call at night?

Record your answer. Then fill in the bubbles. Be sure to use the correct place value.

46. The variables w, x, y, and z each have different whole-number values.

$$w \times y = w$$
$$w \times x = 0$$
$$y + x = z$$

If the value of w is **not** 0, what is the value of z? Explain how you arrived at your answer.

_____
_____
_____
_____

47. There are seven students in Kurt's science class. The table below lists their scores on a recent test.

Student	Score
Jennifer	74
Kathy	82
Jasmine	93
Vishnu	98
Tilly	78
Kurt	95
Lloyd	68

How many points above the class mean is Kurt's score?

Record your answer. Then fill in the bubbles. Be sure to use the correct place value.

48. The sum of the first n positive integers can be determined with the expression $\frac{n(n+1)}{2}$. For example, the sum of the first three integers, 1, 2, and 3, is equal to $\frac{3(3+1)}{2} = \frac{3(4)}{2} = \frac{12}{2} = 6$. How many integers need to be added together to obtain a sum of 55?

49. Jordan works part-time at a restaurant after school and on weekends. The graph below shows how much money he earned in the first four months of the year and how much of that money he spent each month.

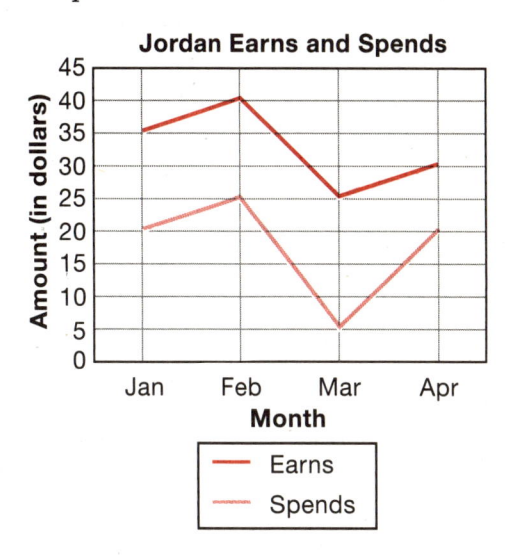

Jordan Earns and Spends

Jordan saves all the money he doesn't spend. In which month did he save the most money? Explain your answer.

_____
_____
_____

50. Susie can input 3 pages in 16 minutes on her computer. At this rate, how many pages could she input in 4 hours?

51. On Saturday, a video game store sold 7 sports games for every 4 role-playing games. The total number of sports and role-playing games sold was 55. How many role-playing games were sold?

Record your answer. Then fill in the bubbles. Be sure to use the correct place value.

52. The function table below illustrates the relationship between the variables x and y.

x	y
-2	4
-1	1
0	0
1	1
2	4

What is the value of y when the value of x is -4?

53. The mean of five integers is 18. If four of the integers are 10, 14, 15, and 17, what is the value of the fifth integer?

54. People in the United States each use an average of 50 gallons of water per day. Estimate the number of gallons of water that a family of 5 would use over the course of two weeks. Explain your answer.

_____
_____
_____

55. Point A is at $(2, -3)$. What are the coordinates of this point after a reflection across the x-axis followed by a reflection across the y-axis?

56. Cordelia made a large cube from 27 small cubes. Then she painted the outside faces of the large cube. What fraction of the small cubes have only one painted face?

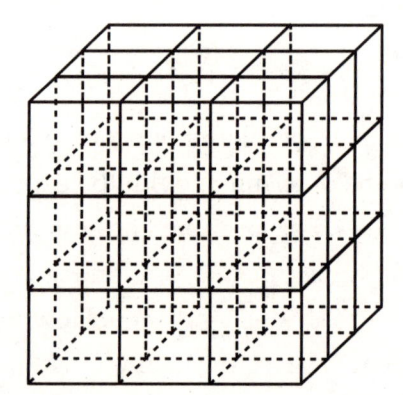